Principles of Genome Analysis

A Guide to Mapping and Sequencing
DNA from Different Organisms

S. B. PRIMROSE

SECOND EDITION

**Blackwell
Science**

19407572

/

© 1995, 1998 by
Blackwell Science Ltd
Editorial Offices:
Osney Mead, Oxford OX2 0EL
25 John Street, London WC1N 2BL
23 Ainslie Place, Edinburgh EH3 6AJ
350 Main Street, Malden
  MA 02148 5018, USA
54 University Street, Carlton
  Victoria 3053, Australia

Other Editorial Offices:
10 rue Casimir Delavigne
75006 Paris, France

Blackwell Wissenschafts-Verlag GmbH
Kurfürstendamm 57
10707 Berlin, Germany

Blackwell Science KK
MG Kodenmacho Building
7–10 Kodenmacho Nihombashi
Chuo-ku, Tokyo 104, Japan

A catalogue record for this title
is available from the British Library

ISBN 0-632-04983-9

Library of Congress
Cataloging-in-publication Data

Primrose, S.B.
    Principles of genome analysis:
  a guide to mapping and sequencing
  DNA from different organisms /
  S.B.Primrose. — 2nd ed.
        p.   cm.
    Includes bibliographical references
    and index.
    ISBN 0-632-04983-9
    1. Gene mapping.   2. Nucleotide
    sequence.   I. Title.
    [DNLM: 1. Chromosome Mapping.
    2. Base Sequence.
    QH 445.2 P953p   1998]
  QH445.2.P75   1998
  572.8'633—dc21
  DNLM/DLC
  for Library of Congress   97-36818
                                   CIP

DISTRIBUTORS

  Marston Book Services Ltd
  PO Box 269
  Abingdon, Oxon OX14 4YN
  (Orders: Tel: 01235 465500
              Fax: 01235 465555)

USA
  Blackwell Science, Inc.
  Commerce Place
  350 Main Street
  Malden, MA 02148 5018
  (Orders: Tel: 800 759 6102
              781 388 8250
              Fax: 781 388 8255)

Canada
  Copp Clark Professional
  200 Adelaide St West, 3rd Floor
  Toronto, Ontario M5H 1W7
  (Orders: Tel: 416 597-1616
              800 815-9417
              Fax: 416 597-1617)

Australia
  Blackwell Science Pty Ltd
  54 University Street
  Carlton, Victoria 3053
  (Orders: Tel: 3 9347 0300
              Fax: 3 9347 5001)

First published 1995
Second edition 1998

Set by Semantic Graphics, Singapore
Printed and bound in Great Britain
at the Alden Press Ltd,
Oxford and Northampton

The Blackwell Science logo is a
trade mark of Blackwell Science Ltd,
registered at the United Kingdom
Trade Marks Registry

Principles of Genome Analysis

Contents

Preface

Since the beginning of this century, a central problem in genetics has been the creation of maps of whole chromosomes. These maps are crucial for the understanding of the structure of genes, their function and their evolution. Until recently these maps were created by genetic means, i.e. as a result of sexual crosses. In the last 10 years the widespread use of recombinant DNA technology has permitted the generation of molecular or physical maps defined here as the ordering of distinguishable DNA fragments by their position along the chromosome. In some instances these physical maps can be as detailed as the complete DNA sequence of entire chromosomes.

Physical mapping of a wide range of genomes has occurred with incredible speed. Already a wide range of manipulative and analytical repertoires exist along with specialist journals and a specialist language. This means that it is very difficult for the experienced geneticist or molecular biologist entering the field to comprehend the latest developments or even what has been achieved. The aim of this text is to provide a general overview of the methodology and rationale employed. In such a fast-moving field it is difficult to be completely up-to-date but for those who worry about such a thing the literature available as late as September 1997 has been surveyed.

A number of individuals provided much appreciated assistance during the preparation of the manuscript and must be acknowledged here. In particular, John Armour and Jay Lewington read the draft text and made many suggestions for improving it. Most of their recommendations were incorporated but any errors or omissions which remain are entirely my responsibility. Thanks also are due to Vera Butterworth and Lynne Goodman for cheerfully finding and checking references at short notice. Without their help the preparation of this edition would have been much more difficult. This list would not be complete without a mention of my family, and my wife in particular, who were expected to provide support and sustenance during the long hours I spent hunched over a keyboard. Hopefully, the final product justifies their efforts and sacrifices!

Abbreviations

ACR	ancient conserved region
AFLP	amplified fragment length polymorphism
AFM	atomic force microscopy
APP	amyloid precursor protein
ARS	autonomously replicating sequence
BAC	bacterial artificial chromosome
BCR	bacterial conserved region
bp	base pair
CAPS	cleaved amplified polymorphic sequences
CCD	charged couple device
CEPH	Centre d'Etude du Polymorphisme Humain
CHEF	contour-clamped homogenous electrical field
cM	centimorgan
ct	chloroplast
DIRVISH	direct visual hybridization
DMD	Duchenne muscular dystrophy
EMC	enzyme mismatch cleavage
ERIC	enterobacterial repetitive intergenic consensus (sequence)
ES	embryonic stem (cells)
EST	expressed sequence tag
FACS	fluorescence-activated cell sorting
FIGE	field-inversion gel electrophoresis
FISH	fluorescence *in situ* hybridization
G6PD	glucose-6-phosphate dehydrogenase
GDRDA	genetically directed representational difference analysis
GMS	genome mismatch scanning
GSS	genome sequence sampling
HAC	human artificial chromosome
HAEC	human artificial episomal chromosome
HPRT	hypoxanthine phosphoribosyl transferase
HTF	*HPa*II Tiny Fragment
kb	kilobase

LINE	long interspersed nuclear element
LOD	logarithm_{10} of odds
LTR	long terminal repeat
Mb	megabase
mt	mitochondrial
OFAGE	orthogonal-field-alteration gel electrophoresis
ORF	open-reading frame
PAC	P1-derived artificial chromosome
PCR	polymerase chain reaction
PFGE	pulsed field gel electrophoresis
QTL	quantitative trait loci
RAPD	randomly amplified polymorphic DNA
RARE	RecA-assisted restriction endonuclease
RC	recombinant congenic (strains)
RDA	representational difference analysis
rDNA/RNA	ribosomal DNA/RNA
REP	repeated extragenic palindrome
RFLP	restriction fragment length polymorphism
RI	recombinant inbread (strains)
SINE	short interspersed nuclear element
STC	sequence-tagged connector
STM	scanning tunnelling microscopy
STS	sequence-tagged site
TAFE	transversely alternating-field electrophoresis
THC	tentative human consensus (sequence)
UTR	untranslated region
YAC	yeast artificial chromosome

1 Rationale for mapping and sequencing genomes

Introduction

Currently a whole series of international efforts is underway to construct genetic and physical maps and to sequence the genomes of a diversity of organisms ranging from the bacterium *Escherichia coli* to humans. The task is a Herculean one. On a global basis it is occupying thousands of scientists and is costing hundreds of millions of dollars annually. Given the effort being expended, the task clearly is a complex one. So, how is the task being approached and why is it being undertaken in the first place? This chapter will focus mainly on the second question, i.e. the rationale for mapping and sequencing genomes. The remainder of the book will concentrate on the methodology for mapping and sequencing genomes and the application of the knowledge gained.

Mapping genes and genomes

Mapping genes and the creation of genetic maps is a fundamental part of the science of genetics. This is because much of genetics is concerned with the understanding and manipulation of the inheritance of particular traits. For plants and animals of agronomic importance this means selective breeding and the identification of those offspring with the desired combination of characteristics. In the case of humans the objective is to predict whether the fetus carries genes for important inherited disorders, i.e. prenatal detection of genetic diseases. Where traits are associated with particular genes the task is relatively easy. Unfortunately there are many times more traits than identified genes and so geneticists make do with marker genes. These are genes which can easily be identified and which are genetically linked to the gene for the trait of interest.

In order to map the locus of a trait by genetic linkage, a panel of markers is tested in turn for evidence of co-segregation with the trait at meiosis. Two loci, A and B, are genetically linked if the alleles present at those loci on a particular chromosome tend to be transmitted together through meiosis. To be linked it is necessary, but not sufficient, for loci to be syntenic, i.e. on the same chromosome. The combination of alleles at linked loci is called a haplotype; for example, haplotype A1B1 means a single chromosome carrying allele A1 at locus A and allele B1 at locus B. During meiosis, each pair of homologous chromosomes undergoes at least one recombination (cross-over) between non-sister chromatids. To show genetic linkage, loci must be located in close physical proximity on a chromosome. Thus, to map a new gene it is necessary to have a large number of different markers, ideally evenly spaced along each chromosome. In micro-organisms such as *E. coli* and yeast, such markers can be generated quite easily by classical mutation techniques. Moving up the evolutionary tree, the generation of markers becomes increasingly difficult and in humans is ethically unacceptable. Instead, naturally occurring polymorphisms are used. By definition, all polymorphisms occur at the level of DNA. Unfortunately few of them are readily scorable like the phenotypes used by Mendel in his classical work on peas. The advent of recombinant DNA technology suggested a completely new approach to defining potentially large numbers of marker loci: the use of DNA probes to identify polymorphic DNA sequences (Botstein *et al.* 1980). The first such DNA polymorphisms to be detected were differences in the length of DNA fragments after digestion with sequence-specific restriction endonucleases, i.e. restriction fragment length polymorphisms (RFLPs) (Fig. 1.1).

To generate an RFLP map the probes must be highly informative. This means that the locus must not only be polymorphic, it must be *very* polymorphic. If enough individuals are studied, any randomly selected probe will eventually discover a polymorphism. However, a polymorphism in which one allele exists in 99.9% of the population and the other in 0.1% is of little utility since it seldom will be informative. Thus, as a general rule, the RFLPs used to construct the genetic map should have two, or perhaps three, alleles with equivalent frequencies.

The first RFLP map of an entire genome (Fig. 1.2) was that described for the human genome by Donis-Keller *et al.* (1987). They tested 1680 clones from a phage library of human genomic DNA to see whether they detected RFLPs by hybridization to Southern blots of DNA from five unrelated individuals. DNA from each individual was digested with 6–9 restriction enzymes. Over 500 probes were identified that detected variable banding patterns indicative of polymorphism. From this collection, a subset of 180 probes detecting the highest degree of polymorphism was selected for inheritance studies in 21 CEPH families (see below). Additional probes were generated from chromosome-specific libraries such that ultimately 393 RFLPs were selected. The various loci were arranged into linkage groups representing the 23 human chromosomes by a combination of mathematical linkage analysis and physical location of selected clones. The latter was achieved by hybridizing probes to panels of rodent–human hybrid cells containing varying human chromosomal complements (see p. 53). RFLP maps have not been restricted to the human genome. For example, RFLP maps have been published for most of the major crops (see for example Moore *et al.*, 1995).

The human genome map produced by Donis-Keller *et al.* (1987) was a landmark publication. However, it identified RFLP loci with an average spacing of 10 centimorgans (cM). That is, the loci had a 10% chance of recombining at meiosis. Given that the human genome is 4000 cM in length, the distance between the RFLPs is 10 Mb on average. This is too great to be of use for gene isolation. However, if the methodology of Donis-Keller *et al.* (1987) was used to construct a 1 cM map, then 100 times the effort would be required! This is because ten times as many probes would be required and ten times more families studied. The solution has been to use more informative polymorphic markers and other mapping techniques and these are described in detail in Chapter 4. Use of these techniques has led to the generation of a

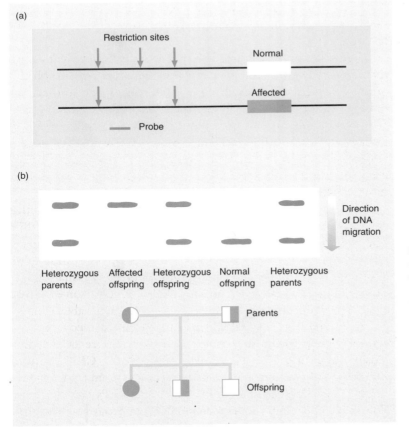

Figure 1.1 Example of a RFLP and its use for gene mapping. (a) A polymorphic restriction site is present in the DNA close to the gene of interest. In the example shown, the polymorphic site is present in normal individuals but absent in affected individuals. (b) Use of the probe shown in Southern blotting experiments with DNA from parents and progeny for the detection of affected offspring.

human map with a marker every 0.7 cM on average (Dib *et al.*
1996) and a mouse map with markers every 0.2 cM (Dietrich *et al.*
1996). More importantly, these advances in gene mapping have led
to increased emphasis in developing representative genetic maps
for a wide range of species, particularly domestic plants and

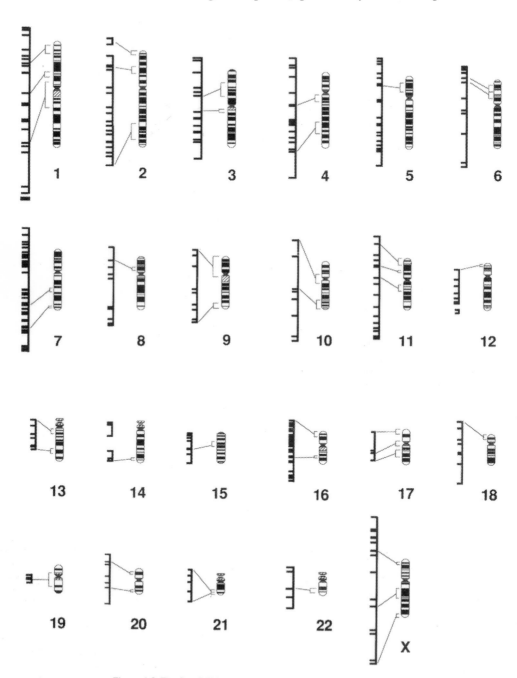

Figure 1.2 The first RFLP genetic linkage map of the entire human genome. (Reproduced
with permission from Donis-Keller *et al.* 1987, © Cell Press.)

animals, e.g. rice (Kurata *et al.* 1994) and pigs (Archibald 1994b).

It should be noted that humans represent an extreme case of difficulty in creating a genetic map. Not only are directed matings not possible, but the length of the breeding cycle (minimum 15–16 years) makes conventional analysis impractical. Consequently, the Centre d'Etude du Polymorphisme Humain (CEPH) was organized in Paris in 1984. The CEPH maintains cell lines from three-generation human families, consisting in most cases of four grandparents, two parents and an average of eight children (Dausset *et al.* 1990). Originally cell lines from 40 families were kept but the number now is much larger. Such families are ideal for genetic mapping because it is possible to infer which allele was inherited from which parent (Fig. 1.3). The CEPH distributes DNAs from these families to collaborating investigators around the world.

Understanding the phenotype

Using classical genetics it is relatively easy to show the mode of inheritance of a particular Mendelian trait, i.e. dominant vs. recessive, autosomal vs. sex linked. A full understanding of the biological basis of the phenotype requires a detailed knowledge of the appropriate gene(s), the genetic control of the gene(s) and identification of the gene product and its role in the life of the host. In some cases the jump from phenotype to gene analysis and biochemical explanation is relatively simple; for example, amino-acid auxotrophy and antibiotic resistance in micro-organisms or phenylketonuria and glucose-6-phosphate dehydrogenase (G6PD) deficiency in humans are phenotypes which are easy to analyse biochemically. However, for the vast majority of traits, including over 4000 in humans, no biochemical information is available. Recent developments in mapping and cloning have provided a solution, originally known as *reverse genetics* but now called

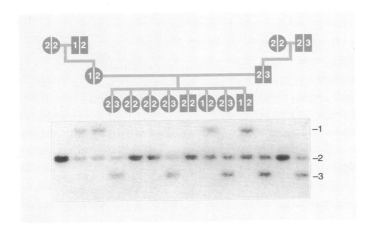

Figure 1.3 Inheritance of a RFLP in a CEPH family. The RFLP probe used detects a single locus on human chromosome 5. In the family shown, three alleles are detected on Southern blotting after digestion with *Taq*I. For each of the parents it can be inferred which allele was inherited from the grandmother and which from the grandfather. For each child the grandparental origin of the two alleles can then be inferred. (Redrawn with permission from Donis-Keller *et al.* 1987, © Cell Press.)

positional cloning. In essence, one maps the gene, clones it, sequences it and then deduces function. The methodology for doing this is described in Chapters 6 and 7. Among the successes achieved with the use of positional cloning, the most significant must be the determination of the biochemical deficiency responsible for cystic fibrosis. The genetics of the disease were well understood (autosomal recessive) and the phenotype (viscid secretions, chest infections, limited lifespan) was easily recognizable. However, until the gene was mapped, cloning and sequencing were not possible. Once sequence analysis could be undertaken it was relatively easy to show that cystic fibrosis is due to deletion of a phenylalanine codon in a gene encoding a chloride transport protein. Increasingly, the function of a gene in one organism, e.g. humans, is elucidated by comparison with a homologous gene in a completely unrelated organism such as yeast.

Comparative genome mapping

As noted above, recent advances in gene mapping have led to the development of representative genetic maps for a wide range of species. Initially these maps were compiled to provide a resource for genetic analysis in the species selected. More recently it has been realized that there is considerable conservation of gene order, not only in related species such as cereals, but in distantly related organisms such as the puffer fish and humans (see Chapter 7). This has two benefits. First, it is possible to transfer linkage information from 'map-rich' species to 'map-poor' species, thereby speeding up map construction. As more and more mapping takes place the increase in our databases should occur exponentially. Second, comparison of the maps should help dissect evolution of genome organization and provide clues about the adaptive rationale, if any, behind particular structural arrangements.

A good example of how molecular mapping helps the understanding of evolutionary processes is provided by an analysis of speciation in monkey flowers (Bradshaw *et al.* 1995). Two *Mimulus* species which are genetically isolated because of a difference in the birds which pollinate them, were found to differ morphologically in eight characteristics. As this organism lacks a detailed genetic map of the conventional kind, a map was constructed using DNA markers. When analysed using these molecular markers, each of the morphological differences was found to involve at least one gene of major effect. Such studies will eventually resolve one of our oldest controversies in genetics: whether adaptations are almost always based on a very large number of mutations, each of small effect, or on single changes of large effect. The advent of DNA markers means that genetic analysis of speciation and adaptation now can be extended to any species of interest rather than being restricted to those which have been intensively studied in the laboratory.

Why sequence genomes?

DNA is the genetic material in all cells. As such, it governs every facet of their existence. The information carried in the DNA determines when and how and where cells grow and divide; for example, why some yeast cells divide by budding and others by binary fission. But what are the differences at the DNA level? And why are some cells prokaryotic and others eukaryotic and what is the difference at the DNA level between simple and complex eukaryotes? Is there a core set of genes for all organisms? Can any organism be considered a 'model organism'. DNA has provided the basis for the evolutionary process that has generated the millions of different life-forms that exist on earth. We are all familiar with the story of Darwin's finches, but what happened at the level of the DNA? The same question can be asked for the monkey flower speciation example given above. Does evolution proceed in the ancestor–descendant pattern that Darwin saw for multicellular plants and animals, or are there other patterns, such as those of lateral gene transfer, obscuring the ancient relationships between domains of life? The answers to these questions, and many others like them, can be inferred from the vast knowledge we have of cellular and molecular biochemistry. However, they can only be answered definitively by studying the sequence of DNA in different organisms.

Although much of our knowledge of cellular and molecular biology was elucidated without DNA sequence information, this was cause rather than effect. Techniques for sequencing DNA did not exist. However, a complete understanding of many biological phenomena was dependent on the development of DNA sequencing techniques. Furthermore, recent experience has shown that the analysis of sequence data is a cost-effective way to generate answers to fundamental questions like those raised above, i.e. sequencing at the beginning of an investigation can be just as worthwhile, if not more so, than at the end. But is it necessary to sequence entire genomes? The answer to this must be in the positive.

Detailed understanding of an organism will only be achieved when every gene has been identified and its transcript and the timing of transcript synthesis known. As a minimum this demands knowledge of the complete gene sequence. In this context it is worth noting that when this was first available for a chromosome, that of yeast chromosome III (Oliver *et al.* 1992), the gene density was much higher than expected. An understanding of evolution will require comparative analysis of entire genomes rather than individual genes; for example, many bacterial genes are organized into operons which are transcribed as polycistronic mRNAs. By contrast, eukaryotic genes were thought to be regulated individually and transcribed as monocistronic mRNAs. Now analysis of 2 megabases (Mb) of DNA sequence from the nematode *Caenorhabditis elegans* has shown that it too has operons, i.e. it uses both the prokaryotic and eukaryotic patterns of gene organization (Zorio *et al*. 1994). Is the *C. elegans* genome becoming more compact to achieve bacterial status or is it expanding towards the eukaryotic monocistronic design?

Now that the complete genomic sequences are available for a number of bacteria and at least one eukaryote (see Chapter 5), it will soon be possible to identify at the DNA level what changes are necessary to go from being a prokaryote to being a eukaryote. Already an analysis of the sequences of the different bacteria is providing information on their physiology and biochemistry as well as their cultural characteristics and eco-pathology. Given that the complete sequence of one of these bacteria (*Mycoplasma genitalium*) was obtained for an estimated cost of only $200 000 then this information was obtained for a fraction of the cost of more conventional studies. The *M. genitalium* genome comprises only about 470 genes and is a big step towards what is the minimal gene complement compatible with a cellular existence. Based on comparisons with other genomic sequences it is possible that an organism might exist with only 200 genes (Koonin *et al*. 1996). With the knowledge acquired to date it might be possible to design experiments specifically focused on the discovery of bacteria with such a tiny genome that they might have escaped detection because of their inability to grow outside a host organism.

Already DNA sequence information is being used to great effect in disciplines such as palaeontology and archaeology (von Haeseler *et al*. 1995); for example, if the demographic history of a population is not known it can be reconstructed from the patterns of nucleotide substitutions in the genome. DNA sequences from the mitochondrial genome and those from the majority of the Y chromosome are particularly useful as they are passed on without recombination from mother to daughter and from father to son. Consequently, these sequences can be traced back directly to the genealogical maternal or paternal 'most recent common ancestor'. Molecular analysis of such sequences suggests that the human species originated in Africa only 100 000 to 200 000 years ago. From there it went on to colonize the world, replacing other human forms such as Neanderthals in the process.

Although the mapping of disease genes in humans, and useful traits in plants and animals, has been undertaken successfully in the past in the absence of sequence information, it now is generally recognized that a genome-wide effort will be more efficient in the long run. This is particularly true when one is seeking to control complex or quantitative traits (see Chapter 7) or to separate the genetic and environmental components of particular diseases. Pharmaceutical companies have taken a great interest in the human genome project for they see the data generated providing an understanding of multifactorial disease. This will enable them to design drugs that will treat the *causes* of disease rather than the *symptoms*, e.g. drugs that act at the level of transcription as opposed to the protein product or, alternatively, oligonucleotides for gene therapy. At the end of the day, the most detailed map available would be one in which every base pair had been identified. This being said, one must question the need to sequence the entire 3000 Mb of human DNA when there is a good chance that at least 2500 Mb of it will be uninformative (but see p. 104). After all, most of the single human gene disorders of any real frequency or medical importance have already been isolated. The real issue is that how much mapping and sequencing is of value depends on the type of genome and the questions one wants the genome map or sequence to answer. For example, it probably would not be worthwhile sequencing the sheep genome which contains a very high proportion of non-coding DNA, homologous with other ruminant genomes, if all we wished to study were a few quantitative traits of commercial importance. By contrast, the yeast genome has been completely sequenced and this was worthwhile: it is very gene dense and it can act as a model for the basic eukaryotic genome.

Genome sequencing projects

A large number of different genomes are being, or have been, sequenced and the most important are listed in Table 1.1. Each was selected for a different reason. There should be no surprise over the choice of *Escherichia coli*. Of all organisms it probably is the best characterized, both genetically and biochemically. *Bacillus subtilis* is of interest, partly because it is Gram-positive whereas *E. coli* is Gram-negative, and partly because it undergoes differentiation during the process of sporulation. The first cellular organism whose genome was completely sequenced was *Haemophilus influenzae* (Fleischmann *et al*. 1995) and it was selected simply because relatively little was known about it, i.e. it acted as a test case. This was followed fairly quickly by the complete sequences of *Mycoplasma genitalium* (Fraser *et al*. 1995), with its near 'minimal' genome, and *Methanococcus jannaschii* (Bult *et al*. 1996) which is a member of the Archaeae. Complete genomic sequencing of many other prokaryotes is being undertaken and the organisms selected are from all the major lineages.

One genome project is focusing on the smallest eukaryotic

genomes identified so far, the nucleomorph genomes from the cryptophytic and chlorarachniophytic algae (McFadden *et al.* 1997). Nucleomorphs are remnant nuclei of former free-living eukaryotic algae that have been engulfed by another eukaryotic cell. The resulting chimaeric cell contains four genomes: the plastid, mitochondrion, host nucleus and nucleomorph (endosymbiotic nucleus). Coevolution between the two partner cells has resulted in the nucleus of the newest host assuming master control of plastid gene expression and the endosymbiotic nucleus has undergone significant reduction in genome size. All the nucleomorphs examined so far contain three small linear chromosomes. In chlorarachniophytes the nucleomorph genome is 380 kb in length and in cryptophytes it is 600 kb in length.

The first eukaryotic genome to be completely sequenced was that of *Saccharomyces cerevisiae* (Goffeau *et al.* 1996). It was selected because, like *E. coli*, it was well characterized genetically. In addition, because it has a nucleus and chromosomes, undergoes meiosis and mitosis etc., analysis of its genome sequence should provide much information on what constitutes a eukaryote. Already this analysis has begun (Oliver 1996a,b, 1997). Unlike most cells, *S. cerevisiae* multiplies by budding, hence the interest in *Schizosaccharomyces pombe* which divides by binary fission. Another reason for interest in *S. pombe* is that it has a genome of similar size

Table 1.1. Organisms whose genomes have been or are being sequenced. Up-to-date information can be found from the TIGR Database on the worldwide web (http://www.tigr.org.org/tdb/)

Genomes completely sequenced
Haemophilus influenzae (Fleischmann *et al.* 1995)
Mycoplasma genitalium (Fraser *et al.* 1995)
Mycoplasma pneumoniae (Himmelreich *et al.* 1996)
Methanococcus jannaschii (Bult *et al.* 1996)
Saccharomyces cerevisiae (Goffeau *et al.* 1996)
Synechocystis sp. (Kaneko *et al.* 1996)
Helicobacter pylori (Tomb *et al.* 1997)
Escherichia coli (Blattner *et al.* 1997)
Bacillus subtilis (in press)
Archaeglobus fulgidus (in press)
Borrelia burgdorferi (in press)
Methanobacterium thermoautotrophicum
Aquifex aeolicus
Treponema pallidum

Genomes being sequenced
Plasmodium falciparum (malarial parasite)
Schizosaccharomyces pombe (fission yeast)
Caenorhabditis elegans (nematode)
Drosophila melanogaster (fruit fly)
Danio rerio (zebra fish)
Fugu rubripes (puffer fish)
Arabidopsis thaliana (thale cress)
Oryza sativa (rice)
Mus musculis (mouse)
Rattus norvegicus (rat)
Homo sapiens (man)

to that of *S. cerevisiae* but is poorly understood at the genetic level. Thus it can be a useful test system for new mapping and sequencing methods. The genomes of many other eukaryotic protists are being sequenced (Fig. 1.4).

Caenorhabditis elegans is a nematode worm and, as such, is a simple multicellular organism. This organism has been extensively analysed genetically but, more important, the line of descent from the zygote is known for every one of the 2000 cells in the adult worms. In this organism it is possible to identify which genes are expressed and when and in what cells as the different cell lineages branch off in the course of determination and differentiation. It will be the first multicellular animal to have its genome completely sequenced (Ahringer 1997). The inclusion of *Drosophila melanogaster* is no surprise. Not only is it well studied genetically, having been a favourite model for Mendelian genetics, but the genetics of morphogenesis and embryogenesis are well understood.

The advantage of using simple organisms such as *Drosophila* and *C. elegans* for the development of sequencing and mapping strategies is that they have genomes that are many times smaller and simpler than those of humans. The existing vertebrate models (e.g. mouse) are the most relevant when studying the structure, function and regulation of mammalian genes. However, they have very large genomes (billions of basepairs). For this reason the genomes of two fishes, the puffer fish (*Fugu rubripes*) and the zebra fish (*Danio rerio*), have been selected for study. In addition, both are

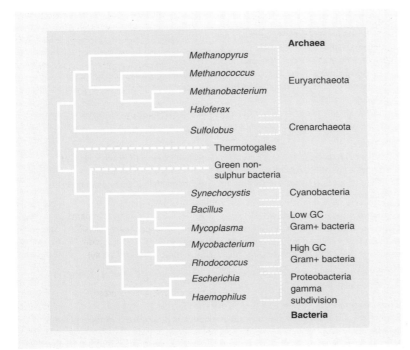

Figure 1.4 The phylogenetic distribution of bacterial species that are the subjects of genome-sequencing projects. The tree topology is based on 16S RNA sequences. The solid lines indicate lineages that are being sequenced. Major branches that are not represented in genome sequencing projects are shown by broken lines. (Redrawn with permission from Koonin *et al.*, 1996.)

vertebrates. The zebra fish is being used as a model for vertebrate embryogenesis (Felsenfeld 1996; Postlethwait & Talbot 1997). It is very easy to isolate mutants of the zebra fish which are affected in normal embryonic development. Because the fundamental molecular mechanisms are shared among vertebrates, the analysis of these mutants is likely to provide new information generally relevant to other classes of vertebrate. *Fugu* has a genome size of 400 Mb and contains less than 10% repetitive DNA. Whilst its genome is 7.5 times smaller than that of humans it has the same amount of coding sequence (Elgar *et al*. 1996). For this reason it has been proposed as a model for vertebrate genome analysis (Brenner *et al*. 1993). It also has been suggested that the *Fugu*

Table 1.2 Goals of the 5-year plan for the US human genome project (from Collins & Galas 1993)

Genetic map
Complete the 2–5 cM human genetic map by 1995
Develop technology to allow non-expert to type families for medical research
Develop markers that can be screened in automated fashion
Develop new mapping technologies

Physical map
Complete map of the human genome at a resolution of 100 kb
Generate clone libraries with improved stability and lower chimaerism

DNA sequencing
Develop efficient approaches to sequencing several megabases of DNA
Develop technology for high throughput sequencing
Build up a sequencing capacity to a collective rate of 50 Mb per year

Gene identification
Develop efficient methods of identifying genes and for placement of known genes on
    physical maps or sequenced DNA

Technology development
Develop evolutionary and revolutionary new technology
Model organisms
Finish an STS map of mouse genome at 300 kb resolution
Finish the sequence of the *E. coli* and *S. cerevisiae* genomes by 1998 (or earlier)
Continue sequencing *C. elegans* and *Drosophila* genomes with the aim of completing
    C. elegans in 1998
Sequence selected segments of mouse DNA side-by-side with corresponding human
    DNA in areas of high biological interest

Informatics
Continue to create, develop and operate databases including effective tools and
    standards for data exchange
Continue to develop better software for mapping, sequencing, data comparison and
    analysis

Ethical, legal and social issues
Continue to identify and define issues and develop policy options
Develop policy options regarding genetic testing services

Training
Continue to encourage training of scientists in interdisciplinary sciences related to
    genome research

Technology transfer
Encourage and enhance technology transfer both into and out of centres of genome
    research

genome could be used as a tool for the identification of disease genes in mammals, provided that significant regions of similarity exist between *Fugu* and mammalian genomes. However, Gilley *et al.* (1997) have found that the genomic organization and comparative gene order of six *Surfeit* genes and two other genes is very different between *Fugu* and mammals. Thus it cannot be assumed that genes that are linked in *Fugu* will also be linked in mammalian genomes.

The interest in humans stems largely from a desire to understand inherited disease and to find ways of diagnosing it, treating it or preventing it. The 5-year goals of the US human genome project have been published (Collins & Galas 1993) and give a good indication of the technological needs and mapping and sequencing objectives of the genome community as a whole. These objectives are summarized in Table 1.2 and it is worth noting that they incorporate sequencing of the genomes of model organisms, all of which have been described above. Progress has been so good that the goals for the next phase of genomics have been proposed (Table 1.3) (Lander 1996). Since directed breeding is not permitted in humans for ethical reasons, the mouse (*Mus musculus*) is a useful model system. In addition, the mouse is an important species for the evaluation of pharmaceuticals, hence the need for a comparative understanding of its genome relative to humans. The interest in genome analysis in the rat comes from its extensive use in physiological research (James & Lindpainter 1997). Such research has generated a wealth of experience and methodological sophistication for the accurate determination of quantitative phenotypes. For commercial reasons, genome maps at least are being constructed for all the major livestock species, particularly with a view to understanding quantitative traits such as size, protein vs. fat content, weight gain, etc.

A number of genome sequencing projects are underway for the major crop species, particularly cereals. Of these, rice is the most important since not only does it have the smallest cereal genome (430 Mb) but it is a staple food of more than half the world's

1. Routine re-sequencing of multi-megabase regions of human and mouse DNA to unravel the polygenic factors underlying human susceptibilities and predispositions
2. Systematic identification of all common variants in human genes
3. Rapid *de novo* sequencing from other organisms to unlock the record of 3.5 billion years of evolution
4. Simultaneous monitoring of the expression of all genes
5. Development of an arsenal of generic tools for manipulating cell circuitry
6. Monitoring the level and modification state of all proteins
7. Preparation of systematic catalogues of all protein interactions
8. Identification of all basic protein shapes
9. Increased attention to ethical, legal and social issues
10. Public education about the benefits and risks associated with genomics.

Table 1.3. Proposed goals for the next phase of genomics (from Lander 1996)

population. The principal objective of the rice genome project is the identification of genes that will provide resistance to the growing ravages of bacterial, fungal and viral diseases as well as to salinity and drought. It also serves as a model monocot plant. *Arabidopsis thaliana* increasingly is being used as a model organism for general plant genome analysis. It is a small, flowering plant which is easily grown, has a rapid life cycle and gives a prolific seed set. These features, plus the fact that it is diploid and can be self- or cross-pollinated, make it attractive for classical genetics. Its small genome (150 Mb) makes it the dicot of choice for a sequencing project.

The fascination of genome sequencing and mapping

Following the development in 1977 of robust and generally applicable sequencing methodologies, good progress was made in sequencing a number of viral genomes, e.g. the 48.5 kb of phage λ (Sanger *et al*. 1982) and the 172 kb Epstein–Barr virus DNA (Baer *et al*. 1984). This generated much excitement and led a decade ago to proposals to sequence the human genome (3×10^9 bp). However, these proposals engendered much controversy that encompassed politics, personalities and scientific judgements of the probable future course of developments in mapping and sequencing. Among the concerns raised were issues such as: 'is the sequencing of the human genome an intellectually appropriate project for biologists? is sequencing the human genome feasible? what benefits might arise from the project? will these benefits justify the cost and are there alternative ways of achieving the same benefits? and will the project compete with other areas of biology for funding and intellectual resources? Today, nobody asks questions like these. The information which has been generated from programmes on genomic mapping and sequencing has provided fascinating insights into many facets of biology and prompted many new areas of investigation.

One organizational change which has occurred in the last few years is in the way in which major sequencing projects are undertaken. Initially, sequencing of a genome was carried out by a large number of laboratories in cooperation, each being assigned a different part of the genome. Role models for this methodology have been yeast, *E. coli* and *B. subtilis*. Now sequencing is being concentrated in major specialist centres where it is undertaken by technicians, thus freeing scientists to tackle the more intellectually interesting questions.

It has to be realized that, to be carried out effectively, genome sequencing must be managed by skilled scientists. It also is repetitive, boring and labour intensive. These two aspects are not compatible and can lead to a high staff turnover rate. However, if the task was completely uninteresting the human genome project would never have got off the ground, far less have spawned the various sequencing projects listed in Table 1.1. The answer to this paradox lies in the complexity of the task. The different genome sizes and architecture in different organisms mean that different mapping strategies are required and a whole series of methods has evolved. Many of these are very elegant but also very complex. As each new method is applied, whole new avenues of research open up. Indeed, so much information has been generated that a number of dedicated journals, e.g. *Nature Genetics* and *Genomics*, have been created. Rather than being a mindless task, genome mapping and sequencing have become intellectually very demanding. For the experienced molecular or cellular biologist entering this field for the first time, never mind the novice, an understanding of the different methods and their interrelationships is nigh on impossible. The objective of this book is to act as a road map (Fig. 1.5). However, the reader does need some basic knowledge, principally a familiarity with hybridization in its different formats, cloning and polymerase chain reaction (PCR) technology. The reader who does not have these should first consult the monograph by Old and Primrose (1994).

Outline of the rest of the book (Fig. 1.5)

As noted above, it is necessary to have an understanding of the organization and structure of the genomes of different organisms so that the methodology used can be optimized. This topic is covered in Chapter 2 and the reader will soon realize the magnitude of the task. The sheer size of the genome of even a bacterium is such that to handle the DNA in the laboratory we need to break it down into smaller pieces which are handled as clones. The methods for doing this are covered in Chapter 3. The process of putting the pieces back together again is mapping and the different methods of mapping and the different markers used are described in Chapter 4. DNA sequencing technology currently is such that only short stretches (~600 bp) can be analysed in a single reaction. Consequently, the genome has to be fragmented and the sequence of each fragment determined and the total sequence re-assembled (Chapter 5). Fortunately, the tools and techniques used for mapping also can be applied to genome sequencing. Analysing the genetic information buried in sequence data is becoming sophisticated and an overview of the methodology used and its application to the first complete genome sequences is covered in Chapter 6. Finally, a combination of mapping and sequencing can be used to isolate genes corresponding to traits of interest and to determine gene function. This is covered in the final chapter.

Rationale for mapping and sequencing genomes

Mapping and sequencing of genomes tells us how organisms differ at the DNA level. However, such studies tell us little about the functionality of the genome. Understanding functionality requires the construction of expression maps. As well as understanding expression patterns in one organism it would be useful to compare expression patterns from different animal and plant models. This would be particularly useful in understanding, for example, brain function. Expression mapping is outwith the scope of this text but the interested reader should consult Strachan *et al*. (1997).

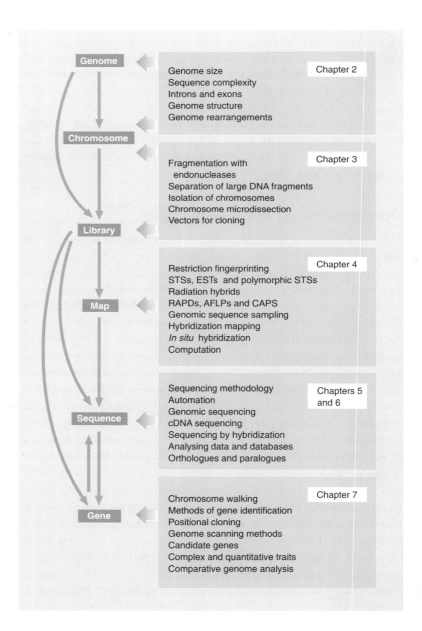

Figure 1.5 'Road map' outlining the different methodologies and interrelationships of genome mapping and sequencing covered in this book.

2 The organization and structure of genomes

Introduction

There is no such thing as a common genome structure. Rather, there are major differences between the genomes of bacteria, viruses and organelles on the one hand and the nuclear genomes of eukaryotes on the other. Within the eukaryotes there are major differences in the types of sequences found, the amounts of DNA and the number of chromosomes. This wide variability means that the mapping and sequencing strategies involved depend on the individual genome being studied.

Genome size

Because the different cells within a single organism can be of different ploidy, e.g. germ cells are usually haploid and somatic cells diploid, genome sizes always relate to the haploid genome. The size of the haploid genome also is known as the C-value. Measured C-values range from 3.5×10^3 bp for the smallest viruses, e.g. coliphage MS2, to 10^{11} bp for some amphibians and plants (Fig. 2.1). The largest viral genomes are $1-2 \times 10^5$ bp and are just a little smaller than the smallest cellular genomes, those of some mycoplasmas (5×10^5 bp). Simple unicellular eukaryotes have a genome size ($1-2 \times 10^7$ bp) that is not much larger than that of the largest bacterial genomes. Primitive multicellular organisms such as nematodes have a genome size about four times larger. Not surprisingly, an examination of the genome sizes of a wide range of organisms has shown that the *minimum* C-value found in a particular phylum is related to the structural and organizational complexity of the members of that phylum. Thus the minimum genome size is greater in organisms that evolutionarily are more complex (Fig. 2.2).

A particularly interesting aspect of the data shown in Fig. 2.1 is the range of genome sizes found within each phylum. Within some phyla, e.g. mammals, there is only a twofold difference between the largest and smallest C-value. Within others, e.g. insects and plants, there is a 10- to 100-fold variation in size. Is there really a 100-fold variation in the number of genes needed to specify different flowering plants? Are some plants really more organizationally complex than humans, as these data imply? Although there is evidence that birds with smaller genomes are better flyers (Hughes & Hughes 1995) and that plants are more responsive to elevated carbon dioxide concentrations (Jasienski & Bazzaz 1995) as their genomes increase in size, this is not sufficient to explain the size differential. The resolution of this apparent C-value paradox has been provided by the analysis of sequence complexity by means of reassociation kinetics.

Sequence complexity

When double-stranded DNA in solution is heated, it denatures ('melts') releasing the complementary single strands. If the solution is cooled quickly, the DNA remains in a single-stranded state. However, if the solution is cooled slowly, reassociation will occur. The conditions for efficient reassociation of DNA were determined originally by Marmur *et al.* (1963) and since then have been extensively studied by others (for a review, see Tijssen 1993). The key parameters are as follows. First, there must be an adequate concentration of cations and below 0.01 M sodium ion there is

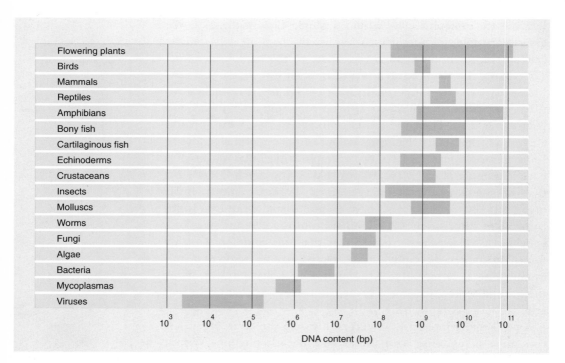

Figure 2.1 The DNA content of the haploid genome of a range of phyla. The range of values within a phylum is indicated by the shaded area. (Redrawn with permission from Lewin 1994, © Cell Press.)

effectively no reassociation. Second, the temperature of incubation must be high enough to weaken intrastrand secondary structure. In practice, the optimum temperature for reassociation is 25°C below the melting temperature (T_m), that is, the temperature required to dissociate 50% of the duplex. Third, the incubation time and the DNA concentration must be sufficient to permit an adequate number of collisions so that the DNA can reassociate. Finally, the size of the DNA fragments also affects the rate of reassociation and is conveniently controlled if the DNA is sheared to small fragments.

The reassociation of a pair of complementary sequences results from their collision and therefore the rate depends on their concentration. Since two strands are involved the process follows second-order kinetics. Thus, if C is the concentration of DNA that is single stranded at time t, then

$$\frac{dC}{dt} = -kC^2$$

where k is the reassociation rate constant. If C_0 is the initial concentration of single-stranded DNA at time $t = 0$, integrating the above equation gives

$$\frac{C}{C_0} = \frac{1}{1 + k.C_0 t}.$$

When the reassociation is half complete, $C/C_0 = 0.5$ and the above equation simplifies to

$$C_0 t_{1/2} = \frac{1}{k}.$$

Thus the greater the $C_0 t_{1/2}$ value, the slower the reaction time at a given DNA concentration. More important, for a given DNA

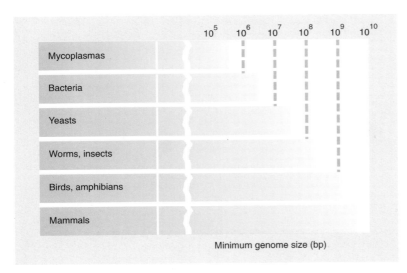

| | 10^5 | 10^6 | 10^7 | 10^8 | 10^9 | 10^{10} |

Mycoplasmas

Bacteria

Yeasts

Worms, insects

Birds, amphibians

Mammals

Minimum genome size (bp)

Figure 2.2 The minimum genome size found in a range of organisms. (Redrawn with permission from Lewin 1994, © Cell Press.)

concentration the half-period for reassociation is proportional to the number of different types of fragments (sequences) present and thus to the genome size (Britten & Kohne 1968). This can best be seen from the data in Table 2.1. Since the rate of reassociation depends on the concentration of complementary sequences, the $C_0t_{1/2}$ for organism B will be 200 times greater than for organism A.

Experimentally it has been shown that the rate of reassociation is indeed dependent on genome size (Fig. 2.3). However, this proportionality is only true in the absence of repeated sequences. When the reassociation of calf thymus DNA was first studied, kinetic analysis indicated the presence of two components (Fig. 2.4). About 40% of the DNA had a $C_0t_{1/2}$ of 0.03, whereas the remaining 60% had a $C_0t_{1/2}$ of 3000. Thus the concentration of DNA sequences that reassociate rapidly is 100 000 times, the concentration of those sequences that reassociate slowly. If the slow fraction is made up of unique sequences, each of which occurs only once in the calf genome, then the sequences of the

Table 2.1 Comparison of sequence copy number for two organisms with different genome sizes

	Organism A	Organism B
Starting DNA concentration (C_0)	10 pg ml^{-1}	10 pg ml^{-1}
Genome size	0.01 pg	2 pg
No. of copies of genome per ml	1000	5
Relative concentration (A vs. B)	200	1

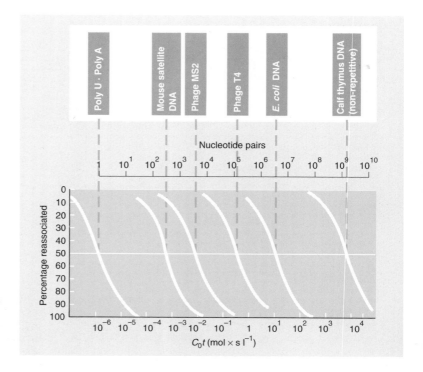

Figure 2.3 Reassociation of double-stranded nucleic acids from various sources. (Redrawn with permission from Lewin 1994, © Cell Press.)

rapid fraction must be repeated 100 000 times, on average. Thus the $C_0t_{1/2}$ value can be used to determine the sequence complexity of a DNA preparation. A comparative analysis of DNA from different sources has shown that repetitive DNA occurs widely in eukaryotes (Davidson & Britten 1973) and that different types of repeat are present. In the example shown in Fig. 2.5 a fast-renaturing and an intermediate-renaturing component can be recognized and are present in different copy numbers (500 000 and 350, respectively) relative to the slow component which is unique or non-repetitive DNA. The complexities of each of these components are 340 bp, 6×10^5 bp and 3×10^8 bp, respectively. The proportion of the genome that is occupied by non-repetitive DNA versus repetitive DNA varies in different organisms (Fig. 2.6), thus resolving the C-value paradox. In general, the length of the non-repetitive DNA component tends to increase as we go up the evolutionary tree to a maximum of 2×10^9 bp in mammals. The fact that many plants and animals have a much higher C-value is a reflection of the presence of large amounts of repetitive DNA. Analysis of mRNA hybridization to DNA shows that most of it anneals to non-repetitive DNA, i.e. most genes are present in non-repetitive DNA. Thus genetic complexity is proportional to the content of non-repetitive DNA and not to genome size. There is, however, a surprising constancy in gene numbers. In all eukaryotes there are 7000–20 000 genes except for the vertebrates which have 50 000–100 000 genes (Bird 1995).

Introns and exons

Introns were initially discovered in the chicken ovalbumin and rabbit and mouse β-globin genes (Breatnach *et al.* 1977; Jeffreys & Flavell 1977). Both these genes had been cloned by isolating the

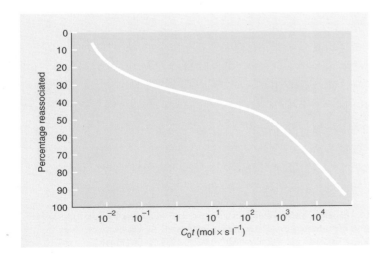

Figure 2.4 The kinetics of reassociation of calf thymus DNA. Compare the shape of the curve with those shown in Fig. 2.3.

mRNA from expressing cells and converting it to cDNA. The next step was to use the cloned cDNA to investigate possible differences in the structure of the gene from expressing and non-expressing cells. Here the Southern blot hybridizations revealed a totally unanticipated situation. It was expected that the analysis of genomic restriction fragments generated by enzymes that did not cut the cDNA would reveal only a single band corresponding to the entire gene. Instead several bands were detected in the hybridized blots. The data could be explained only by assuming the existence of interruptions in the middle of the protein-coding sequences. Furthermore, these insertions appeared to be present in both expressing and non-expressing cells. The gene insertions that are not translated into protein were termed *introns* and the sequences that are translated were called *exons*.

Since the original discovery of introns, a large number of split genes have been identified in a wide variety of organisms. These introns are not restricted to protein-coding genes for they have been found in rRNA and tRNA genes as well. Split genes are rare in prokaryotes and not particularly common in lower eukaryotes.

In *Saccharomyces cerevisiae*, sequencing of the complete genome suggests that there are 235 introns compared with over 6000

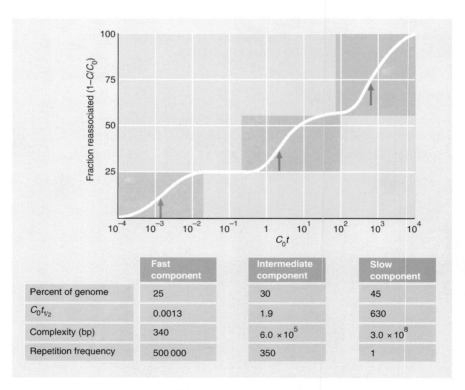

	Fast component	Intermediate component	Slow component
Percent of genome	25	30	45
$C_0 t_{1/2}$	0.0013	1.9	630
Complexity (bp)	340	6.0×10^5	3.0×10^8
Repetition frequency	500 000	350	1

Figure 2.5 The reassociation kinetics of a eukaryotic DNA sample showing the presence of two types of repeated DNA. The arrows indicate the $C_0 t_{1/2}$ values for the three components. (Redrawn with permission from Lewin 1994, © Cell Press.)

open-reading frames and that introns account for less than 1% of the genome (Goffeau *et al*. 1996). Those genes which do have introns usually have only one small one and the longest intron is only 1 kb in size.

However, proceeding up the evolutionary tree, the number of split genes, and the number and size of introns per gene, increases (Fig. 2.7 and Table 2.2). More important, genes that are related by evolution have exons of similar size, i.e. the introns are in the same position. However, the introns may vary in length, giving rise to variation in the length of the genes (Fig. 2.8). Note also that introns are much longer than exons, particularly in higher eukaryotes.

If a split gene has been cloned, it is possible to sub-clone either the exon or the intron sequences. If these sub-clones are used as probes in genomic Southern blots, it is possible to determine if these same sequences are present elsewhere in the genome. Often,

Species	Average exon number	Average intron number	Average length (kb)	Average mRNA length (kb)	% Exon per gene
Yeast	1	0	1.6	1.6	100
Nematode	4	3	4.0	3.0	75
Fruit fly	4	3	11.3	2.7	24
Chicken	9	8	13.9	2.4	17
Mammals	7	6	16.6	2.2	13

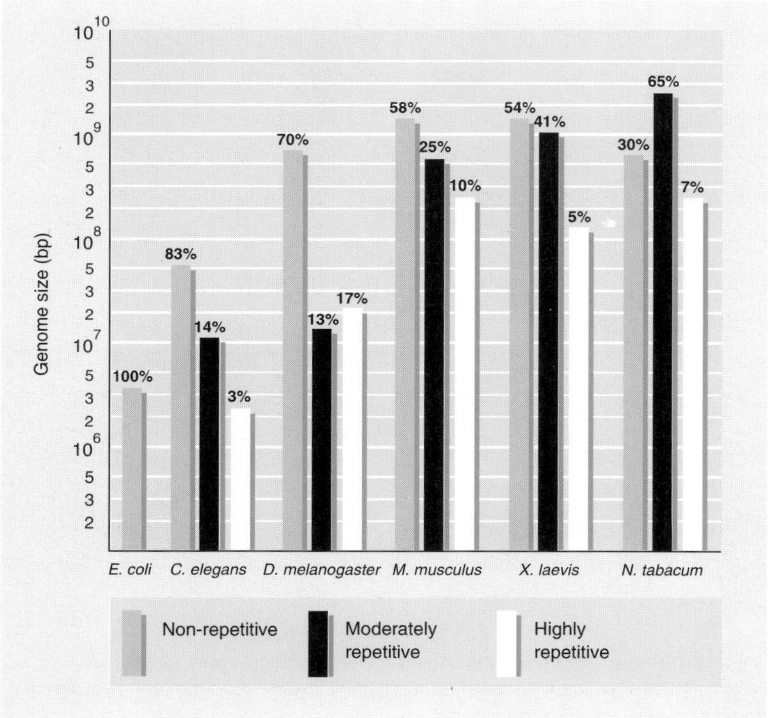

Table 2.2 Intron statistics for genes from different animal species

Figure 2.6 The proportions of different sequence components in representative eukaryotic genomes. (Redrawn with permission from Lewin 1994, © Cell Press.)

the exon sequences of one gene are found to be related to sequences in one or more other genes. Some examples of such *gene families* are given in Table 2.3. In some instances the duplicated genes are clustered, whereas in others they are dispersed. Also, the members may have related, or even identical, functions, although they may be expressed at different times or in different cell types. Thus different globin proteins are found in embryonic and adult red blood cells, while different actins are found in muscle and non-muscle cells. Functional divergence between members of a multigene family may extend to the loss of gene function by some members. Such *pseudogenes* come in two types. In the first type they retain the usual intron and exon structure but are functionless or they lack one or more exons. In the second type, found in dispersed gene families, processed pseudogenes are found which lack any sequences corresponding to the introns or promoters of the functional gene members. Multiple copies of an exon also may be found because the same exons occur in several apparently unrelated genes. Exons that are shared by several genes are likely to

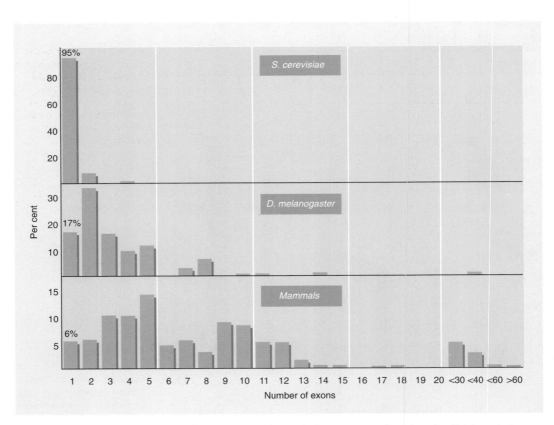

Figure 2.7 The number of exons in three representative eukaryotes. Uninterrupted genes have only one exon and are totalled in the left-hand column. (Redrawn with permission from Lewin 1994, © Cell Press.)

encode polypeptide regions that endow the disparate proteins with related properties, e.g. ATP or DNA binding. Some genes appear to be mosaics that were constructed by patching together copies of individual exons recruited from different genes, a phenomenon known as *exon shuffling*.

By contrast with exons, introns are not related to other sequences in the genome, although they contain the majority of

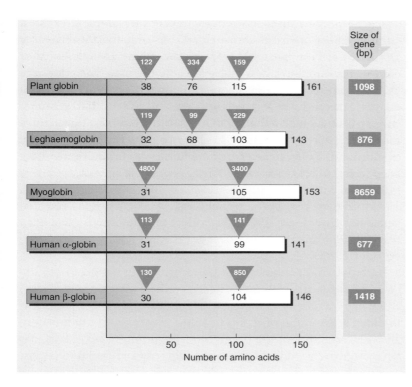

Figure 2.8 The placement of introns in different members of the globin superfamily. The size of the introns in base pairs is indicated inside the inverted triangles. Note that the size of each polypeptide and the location of the different introns are relatively consistent.

Table 2.3 Some examples of multigene families

Gene family	Organism	Approximate no. of genes	Clustered (L) or dispersed (D)
Actin	Yeast	1	—
	Slime mould	17	L, D
	Drosophila	6	D
	Chicken	8–10	D
	Human	20–30	D
Tubulin	Yeast	3	D
	Trypanosome	≈30	L
	Sea urchin	15	L,D
	Mammals	25	D
α-Amylase	Mouse	3	L
	Rat	9	?
	Barley	7	?
β-Globin	Human	6	L
	Lemur	4	L
	Mouse	7	L
	Chicken	4	L

dispersed, highly repetitive sequences. Thus, for some genes the exons constitute slightly repetitive sequences embedded in a unique context of introns. It should be noted that introns are not necessarily junk because there now is evidence that some of them encode functional RNA (Moore 1996).

Genome structure in viruses and prokaryotes

The genomes of viruses and prokaryotes are very simple structures, although those of viruses show remarkable diversity (for review see Dimmock & Primrose 1994). Most viruses have a single linear or circular genome but a few, such as the reoviruses, bacteriophage Φ6 and some plant viruses, have segmented RNA genomes. Until relatively recently it was believed that all eubacterial genomes consisted of a single circular chromosome. However, linear chromosomes have been found in *Borrelia burgdorfii*, *Streptomyces lividans* and *Rhodococcus fascians*, while two chromosomes have been found in *Rhodobacter spheroides*, *Brucella melitensis*, *Leptospira interrogans* and *Agrobacterium tumefaciens* (Cole & Saint Girons 1994). Where two chromosomes have been detected, essential genes for growth are found on both. Linear plasmids have been found in *Borrelia* spp. and *Streptomyces* spp. (Kinashi *et al.* 1987; Barbour 1993).

Bacterial viral genomes lack the centromeres found in eukaryotic chromosomes although there may be a partitioning system based on membrane adherence. Duplication of the genomes is initiated at an origin of replication and may proceed unidirectionally. The structure of the origin of replication, the *ori*C locus, has been extensively studied in a range of bacteria and found to consist essentially of the same group of genes in a nearly identical order (Cole & Saint Girons 1994). The *ori*C locus is defined as a region harbouring the *dna*A (DNA initiation) or *gyr*B (B subunit of DNA gyrase) genes linked to a ribosomal RNA operon.

Many bacterial and viral genomes are either circular or can adopt a circular conformation for the purposes of replication. As such, these molecules do not have telomeres (see p. 31). However, many viral genomes always retain a linear configuration, as do linear plasmid molecules. All these molecules exhibit specialized structural features to facilitate duplication of the ends of the molecules and to protect them from exonuclease digestion. For example, poxviruses have closed hairpin loops at the end of their genomes and phages T2 and T7 exhibit terminal redundancy, i.e. they have the sequence at both ends of the viral DNA. Many other variations have been noted (for review, see Dimmock & Primrose 1994). All the linear chromosomes and plasmids of *Streptomyces* that have been characterized contain terminal inverted repeats and covalently bound proteins. In this respect they resemble the adenoviruses and bacteriophage Φ29. However the mechanisms whereby the ends are replicated are probably different for the viruses and the bacteria (for review, see Chen 1996).

In the original studies on the kinetic analysis of the reassociation of phage and bacterial DNA no repeated sequences were noted (see p. 23). As a further check for repetitive DNA in *E. coli*, Britten & Kohne (1968) isolated the first small fraction to reassociate and compared its reassociation kinetics with that of bulk *E. coli* DNA: no differences were found. While the sensitivity of this test is high, the existence of a small amount of repetitive DNA cannot be ruled out. Indeed, repetitious DNA has been detected. For example, classical genetic studies have revealed the presence of multiple copies of mobile elements known as insertion sequences in the genome of *E. coli* and its relatives. However, the copy number is very small, usually less than 20 per genome, and they represent a very small proportion of the total cellular DNA. Sequencing of the *E. coli* genome has revealed the presence of other repeated sequences such as REP (repeated extragenic palindrome) and ERIC (enterobacterial repetitive intergenic consensus) elements. For example, a 225 kb region of the *E. coli* genome was found to contain 19 REP elements (Sofia *et al.* 1994). Repeated sequences are also found in other bacterial genomes but these are unrelated to REPs and ERICs. Three families of repeated elements have been found in *Methanococcus jannaschii* following analysis of the complete genome sequence (Bult *et al.* 1996) and it is expected that many others will be found as more genomes are sequenced. Also in *E. coli*, a large number of pairs or groups of redundant genes have been identified (Riley 1993) in addition to multiple genes for tRNAs. These redundant genes have similar nucleotide sequences and the gene products similar amino acid compositions.

Split genes appear to be rare in prokaryotes and viruses. Introns have been detected in only a few bacteriophages and in a tRNA*ser* gene in the archaebacterium *Acanthamoeba* (Belfort 1989).

The organization of organelle genomes

Mitochondria and chloroplasts both possess DNA genomes that code for all of the RNA species and some of the proteins involved in the functions of the organelle. In some lower eukaryotes the mitochondrial (mt) DNA is linear but more usually organelle genomes take the form of a single circular molecule of DNA. Because each organelle contains several copies of the genome and because there are multiple organelles per cell, organelle DNA constitutes a repetitive sequence. Whereas chloroplast (ct) DNA falls in the range 120–200 kb, mtDNA varies enormously in size. In animals it is relatively small, usually less than 20 kb. But yeast mtDNA is about 80 kb and that of plants ranges from several hundred to several thousand kilobase pairs. In addition, certain protozoans, e.g. trypanosomes, have within their single mitochondrion a disk structure known as a kinetoplast. This contains thousands of interlocked circular DNA molecules. In *Trypanosoma brucei* about 45 of the circles are 21 kb long and the remaining 5500 are 1 kb long. Considerable sequence variation occurs within the circles (Lamond 1988).

ORGANIZATION OF THE CHLOROPLAST GENOME

The complete sequence of ctDNA has been reported for a liverwort (a bryophyte; Ohyama *et al.* 1986), tobacco (a dicot; Shinozaki *et al.* 1986) and rice (a monocot; Hiratsuka *et al.* 1989) and the general organization of many other ctDNA molecules has been elucidated by restriction endonuclease mapping (see later). A general feature of ctDNA from both higher and lower eukaryotes is a 10–24 kb sequence that is present in two identical copies as an inverted repeat (Fig. 2.9). The only exceptions to this structure are

Table 2.4 Key features of ct DNA

Feature	Liverwort	Tobacco	Rice
Inverted repeats	10 058 bp	25 339 bp	20 799 bp
Short unique sequence	19 813 bp	18 482 bp	
Long unique sequence	81 095 bp	86 684 bp	
Length of total genome	121 024 bp	155 844 bp	134 525 bp
Number of genes	128	84	
Number of genes with introns	20	15	18

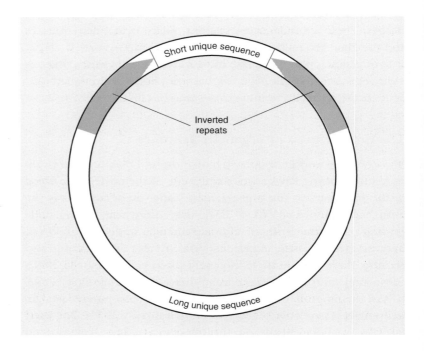

Figure 2.9 Generalized structure of ct DNA.

CHAPTER 2
*The organization and
structure of genomes*

in some legumes that have lost one copy of the inverted repeat and *Euglena* which lacks both copies. However, *Euglena* does have three tandem repeats of a 5.6 kb sequence. Key features of the ctDNA genome are shown in Table 2.4.

ORGANIZATION OF THE MITOCHONDRIAL GENOME

Mitochondrial DNA (mtDNA) varies widely in size and ranges from 16–20 kb in animals, 80 kb in *Saccharomyces cerevisiae* and from 200–2500 kb in higher plants. The largest plant mitochondrial genome is almost half the size of the *E. coli* genome. The 16.6 kb mtDNAs from three species (man, mouse and cow) have been sequenced and the genome organization is similar. There are no introns, some genes overlap and almost every base pair can be assigned to a gene. The mtDNA from *S. cerevisiae* has not been sequenced but about 24% of the genome is known to consist of short stretches of A-T rich DNA and introns are known to occur. The mitochondrial genome of the flagellate *Reclinomonas americana* has been sequenced and found to resemble closely the eubacterial genome (Palmer 1997).

The organization of mtDNA in higher plants is complex, particularly since it can undergo homologous recombination that results in considerable variation within and between species. Closely related species have quite different mitochondrial genome sizes and hence such size variation must have occurred very rapidly in evolutionary terms. Each species is characterized by the presence of repeated sequences, pseudogenes, chimaeric genes and chloroplast DNA sequence insertions. Based on mapping studies the total genetic information of plant mtDNA can be arranged into a single, circular molecule that is known as the master chromosome. This circular DNA molecule contains repeated sequences that can generate, via intramolecular recombination, either isomeric forms of the master chromosome or smaller sub-genomic circular DNA molecules. Fauron *et al* (1995) have proposed a general model which explains the origin of this mtDNA sequence diversity.

Analysis of the mitochondrial genome of *Arabidopsis thaliana* has shown that known genes account for only 10% of the entire sequence (Unseld *et al*. 1997). Introns in these genes, other open-reading frames, duplications, remnants of retrotransposons and integrated plastid sequences account for another 30% of the sequence. However, the biological significance of the remaining 60% of the mitochondrial genome is not known. Also of interest is the observation that the gene density in the mitochondrial genome of *A. thaliana* is half that of the nuclear DNA. Also, there are fewer genes in the 367 kb mitochondrial DNA of *Arabidopsis*

than there are in the 187 kb mtDNA of the liverwort *Marchantia
polymorpha.*

The organization of nuclear DNA in eukaryotes

GROSS ANATOMY

Each eukaryotic nucleus encloses a fixed number of chromosomes
which contain the nuclear DNA. During most of a cell's life, its
chromosomes exist in a highly extended linear form. Prior to cell
division, however, they condense into much more compact bodies
which can be examined microscopically after staining. The duplication of chromosomes occurs chiefly when they are in the
extended stage (interphase). One part of the chromosome, however, always duplicates during the contracted metaphase state. This
is the *centromere*, a body that controls the movement of the
chromosome during mitosis. Its structure is discussed later (p. 42).

In many eukaryotes, a variety of treatments will cause chromosomes in dividing cells to appear as a series of light- and dark-
staining bands (Fig. 2.10). In G-banding, for example, the
chromosomes are subjected to controlled digestion with trypsin
before Giemsa staining which reveals alternating positively (dark
G-bands) and negatively (R-bands or pale G-bands) staining regions. As many as 2000 light and dark bands can be seen along
some mammalian chromosomes. An identical banding pattern
(Q-banding) can be seen if the Giemsa stain is replaced with a
fluorescent dye such as quinacrine which intercalates between the
bases of DNA. A structural basis for metaphase bands has been
proposed that is based on the differential size and packing of DNA
loops and matrix-attachment sites in G- versus R-bands (Saitoh &
Laemmli 1994). Bands are classified according to their relative
location on the short arm (p) or the long arm (q) of specific

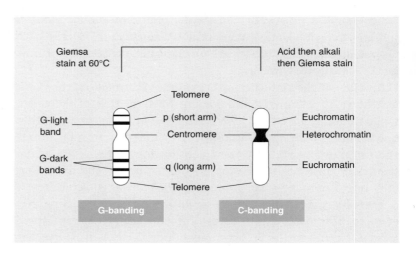

Figure 2.10 Banding patterns
revealed on chromosomes by
different staining methods.
Note that intercalating
fluorescent dyes produce the
same pattern as Giemsa stain
at 60°C. In the C-banding
technique some
heterochromatin may be
detected at the telomeres.

chromosomes; for example, 12q1 means band 1 on the long arm of chromosome 12. If the chromosome DNA is treated with acid and then alkali prior to Giemsa staining, then only the centromeric region stains and this is referred to as *heterochromatin*. The unstained parts of the chromosome are called *euchromatin*.

Because Giemsa shows preferential binding to DNA rich in AT base pairs, the dark G-bands are believed to be A + T rich and light G-bands G + C rich. At the cytogenetic level a clear relationship exists between the G- and R-bands and the organization of genes. R-bands are gene rich and in a more open chromatin configuration, whereas G-bands are gene poor. However, at higher resolution (10–100 kb) R-bands are heterogeneous and the structural and functional relevance of this observation is not clear (Gardiner 1995). It should be noted that DNA which is not transcriptionally active tends to be methylated. Because methylation of restriction enzyme recognition sequences can prevent endonuclease cleavage this can affect the recovery of DNA fragments.

The ends of eukaryotic chromosomes are also the ends of linear duplex DNA and are known as *telomeres*. That these must have a special structure has been known for a long time. For example, if breaks in DNA duplexes are not rapidly repaired by ligation they undergo recombination or exonuclease digestion, yet, the ends of chromosomes are stable and chromosomes are not ligated together. Also, DNA replication is initiated in a 5′→3′ direction with the aid of an RNA primer. After removal of this primer there is no way of completing the 5′ end of the molecule (Fig. 2.11). Thus, in the absence of a method for completing the ends of the molecules, chromosomes would become shorter after each cell division.

The determination of the structure of telomeres from a number of organisms has shown that they are not blunt-ended but do exist with single-stranded 3′ overhangs of 12 nucleotides or more. They are maintained in this state, without chromosome shortening, by telomerase, a ribonucleoprotein enzyme that extends DNA termini by virtue of an internal RNA template. Repeated DNA sequences (see p. 37) also are involved. The exact mechanism of

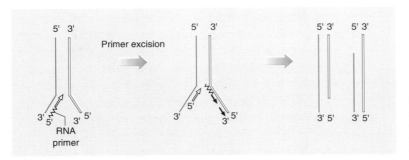

Figure 2.11 Formation of two daughter molecules with complementary single-stranded 3′ tails after primer excision.

telomere maintenance is not known but a number of models have been proposed (Lingner *et al.* 1995). A decrease in size of telomeric DNA may be responsible for cellular death (Holliday 1996).

ORIGINS OF REPLICATION

All of the cell's DNA must be replicated once, and once only, during the cell cycle. In yeast (*S. cerevisiae*) the origins of DNA replication have been identified as ARS (autonomously replicating sequence) elements. Incorporation of the latter on plasmids permits their replication in yeast as yeast artificial chromosomes (YACs) (see p. 55). The sequence requirements for an ARS element are known (Marahrens & Stillman 1992). So far, origins of DNA replication have not been unambiguously identified in other eukaryotes.

REPEATED SEQUENCES

It will be recalled from the section on reassociation kinetics of DNA that repeated sequences are a common feature of eukaryotic DNA. How are these repeated sequences organized in the genome? Are they tandemly repeated or are they dispersed? The answer is that both organizational patterns are found. In general, highly repetitive DNA is organized around centromeres and telomeres in the form of tandem repeats, whereas moderately repetitive DNA is dispersed throughout the chromosome.

Tandemly repeated sequences

Repeated sequences were first discovered 30 years ago during studies on the behaviour of DNA in centrifugal fields. When DNA is centrifuged to equilibrium in solutions of CsCl, it forms a band at the position corresponding to its own buoyant density. This in turn depends on its percentage G + C content:

$$\rho(\text{density}) = 1.660 + 0.00098\ (\%\ GC)\ g\,cm^{-3}.$$

When eukaryotic DNA is centrifuged in this way the bulk of the DNA forms a single, rather broad band centred on the buoyant density which corresponds to the average G : C content of the genome. Frequently one or more minor or *satellite* bands are seen (Fig. 2.12). The behaviour of satellite DNA on density gradient centrifugation frequently is anomalous. When the base composition of a satellite is determined by chemical means it often is different from what had been predicted from its buoyant density. One reason is that it is methylated which changes its buoyant density.

Once isolated, satellite DNA can be radioactively labelled *in*

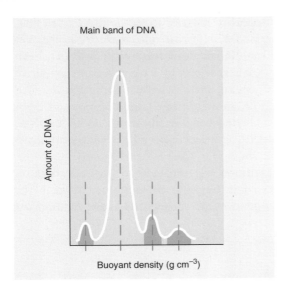

Figure 2.12 Detection of three satellite DNA bands (dark shading) on equilibrium density gradient centrifugation of total DNA.

vitro and used as a probe to determine where on the chromosome it will hybridize. In this technique, known as *in situ hybridization*, the chromosomal DNA is denatured by treating cells that have been squashed on a cover slip. The localization of the sites of hybridization is determined by autoradiography. Using this technique, most of the labelled satellite DNA is found to hybridize to the heterochromatin present around the centromeres and telomeres. Since RNA that is homologous to satellites is found only rarely the heterochromatic DNA most probably is non-coding.

When satellite DNA is subjected to restriction endonuclease digestion only one or a few distinct low-molecular-weight bands are observed following electrophoresis. These distinct bands are a tell-tale sign of tandemly repeated sequences. The reason (Fig. 2.13) is that if a site for a particular restriction endonuclease occurs in each repeat of a repetitious tandem array, then the array is digested to unit-sized fragments by that enzyme. After elution of the DNA band from the gel it can be used for sequence analysis either directly or after cloning. However, the sequence obtained is a consensus sequence and not necessarily the sequence of any particular repeat unit because sequence divergence can and does occur very readily. Note that if such sequence divergence occurs within a restriction endonuclease cleavage site in the repeated units then digestion with the enzyme produces multimers of the repeat unit ('higher order repeats') (Fig. 2.13).

The amount of satellite DNA and its sequence varies widely between species and can be highly polymorphic within a species. Thus 1–3% of the genome of the rat (*Rattus norvegicus*) is centromeric satellite DNA, whereas it is 8% in the mouse (*Mus musculus*) and greater than 23% in the cow (*Bovis domesticus*). The length of the satellite repeat unit varies from the d(AT)$_n$: d(TA)$_n$ structure found in the land crab (*Gecarcinus lateratis*) to the very complicated structure seen in the domestic cow (Fig. 2.14). Even within a single genus each species can have a distinctive set of satellite sequences. In general, little is known about the detailed structure of repeated DNA arrays but one family of repeated DNA that has been analysed extensively is ribosomal DNA (rDNA) (Williams & Robbins 1992). An example of rDNA repeat structure is shown in Fig. 2.15.

Not all tandem repetitions are restricted to heterochromatin: some are found dispersed throughout the genome, often in the spacer region between genes. One such group is the *minisatellite*

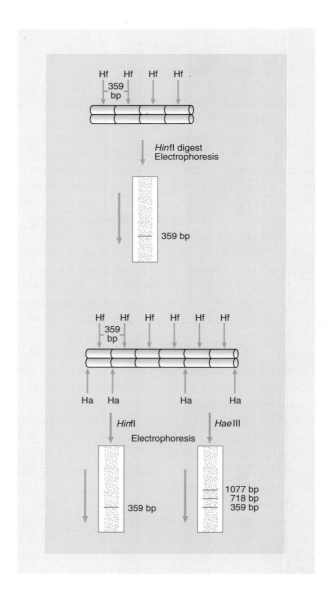

Figure 2.13 Digestion of purified satellite DNA by restriction endonucleases. The basic repeat unit is 359 bp long and contains one endonuclease *Hin*fI (Hf) site. Digestion with *Hin*fI converts most of the satellite DNA to a set of 359 bp long fragments. These are abundant enough to be seen as a band against the smear of other genomic fragments after gel electrophoresis and staining with ethidium bromide. Digestion of the DNA with endonuclease *Hae*III (Ha) yields a ladder of fragments that are multiples of 359 bp in length. (Redrawn with permission from Singer & Berg 1990.)

sequences, which also demonstrates intraspecies polymorphism. This polymorphism has been used for DNA fingerprinting of organisms. When cloned probes containing a minisatellite sequence are annealed with DNA blots containing restriction endonuclease digests of DNA, multiple bands hybridize. The pattern of bands varies from one individual of a species to another but is the same when DNA from several tissues of a single individual is examined. The bands are inherited in a Mendelian fashion and it is possible to identify those bands inherited from each parent (Fig. 2.16). For this reason the technique has forensic applications (for reviews see Monckton & Jeffreys 1993; Alford & Caskey 1994).

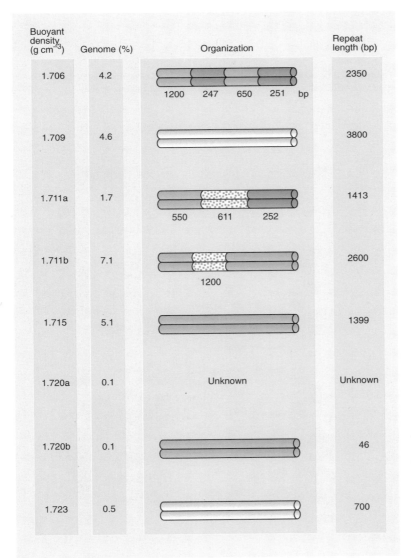

Figure 2.14 The structure of the different types of satellite DNA found in the domestic cow. Homologous portions of the different satellites are indicated by similar colouring. The designations a and b indicate two versions with the same buoyant density. (Redrawn with permission from Singer & Berg 1990.)

Microsatellite DNA families include small arrays of tandem repeats which are simple in sequence (1–4 bp) and which are interspersed throughout the genome. In mammalian cells, runs of $(dA.dT)_n$ are very common and can account for 0.3% of the genome. In contrast, runs of $(dG.dC)_n$ are much rarer. Runs of dinucleotides are also found; for example, in humans, runs of $(dCA.dTG)_n$ and $(dCT.dAG)_n$ account for 0.5% and 0.2% of the genome respectively. Trinucleotide and tetranucleotide repeats also occur but are rarer. The significance of these repeats in normal

Figure 2.15 Detailed architecture of a rDNA repeat unit. Transcribed regions are shown in boxes. Coding sequences are shown in green and spacer regions in white.

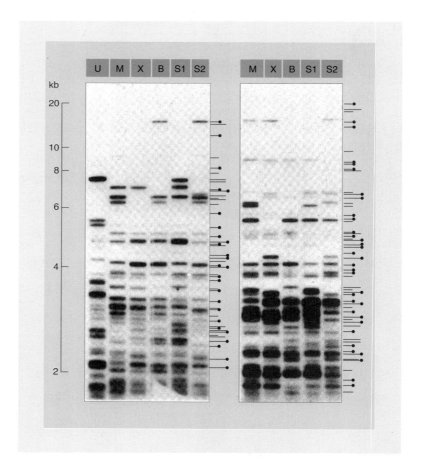

Figure 2.16 Use of minisatellite sequences to detect polymorphisms in human DNA for forensic purposes.

genes is not known but they can be the locus for a number of inherited disorders when they undergo unstable expansion. For example, in fragile X syndrome patients can exhibit hundreds or even thousands of the CGG triplet at a particular site, whereas unaffected individuals only have about 30 repeats. So far, over a dozen examples of disease resulting from trinucleotide expansion have been described (Sutherland & Richards 1995; Warren 1996; Mitas 1977). Similar trinucleotide repeats have been discovered in bacteria (Hancock 1996) and yeast (Dujon 1996) following complete genome sequencing. As in the human case, perfect trinucleotide repeats in yeast are subject to polymorphic size variation while imperfect ones are not. Some pathogenic bacteria use length variation in simple repeats to change the antigens on their surfaces so that they can evade host immune attack.

Repeated sequences are also found in telomeres and those from lower and higher eukaryotes are constructed on the same basic principle. Each telomere consists of a long series of short, tandemly repeated sequences (Table 2.5). All can be written in the general form $C_n(A/T)_m$, where $n > 1$ and m is 1–4. The number of repeats varies greatly but the overall telomere length can be up to 12 kb. The detailed structure of telomeres has been reviewed by Shippen (1993) and Kipling (1995).

Tandem repetition of DNA sequences also occurs within coding regions. For example, linked groups of identical or near-identical genes sometimes are repeated in tandem. These are the gene families described earlier (p. 24). However, tandem repetition also occurs within a single gene; for example, the *Drosophila* 'glue' protein gene contains 19 direct tandem repeats of a sequence 21 base pairs long that encodes seven amino acids. The repeats are not perfect but show divergence from a consensus sequence. Another example is the gene for α2(1) collagen found in chicken, mouse and humans. The gene comprises 52 exons with introns varying in length from 80 to 2000 bp. However, all the exon sequences are multiples of 9 bp and most of them are 54 or 108 bp long. This accounts for the observed primary sequence of collagen which has glycine in every third position and a very high content of proline and lysine.

Table 2.5 Representative example of the repeating units found in telomeres

Organism	Repeating unit (5'→3')
Tetrahymena (ciliate)	CCCCAA
Trypanosoma (flagellate)	CCCTA
Dictyostelium (slime mould)	CCCTA
Saccharomyces (yeast)	$C_{2-3}A(CA)_{1-3}$
Arabidopsis (thale cress)	C_3TA_3
Homo sapiens (human)	C_3TA_2

Dispersed repeated sequences

Moderately repetitive DNA is characterized by being dispersed throughout the genome and this was discovered by an extension of the early work on reannealing kinetics. It will be recalled from p. 19 that reannealing of DNA is dependent on DNA fragment size and that it is common practice to shear the DNA before hybridization begins. The reason for this is that it was observed that most high-molecular-weight fragments would reassociate with each other as a result of the interaction of repetitive DNA sequences. Very large particles, termed networks, were formed. This observation indicated that many fragments as long as 10 000 nucleotides contained more than one repetitive sequence element. When the fragment size was reduced, a smaller number of fragments reassociated at low $C_0t_{1/2}$ values. When the percentage reassociation is determined for different lengths of DNA at C_0t values where single-copy DNA cannot anneal, it is possible to determine the length of the interspersed sequences (Fig. 2.17). From the example shown it can be seen that in human DNA 2.2 kb of relatively rare sequences separate the repeated sequences in 60% of the genome. In *Xenopus* about 800 bp of unique sequences separate repeats in about 65% of the genome. By contrast, in *Drosophila* about 12 kbp separate the repeated sequences. Kinetic analysis of this type has revealed that the predominant interspersion pattern of insects and fungi is long range (i.e. like *Drosophila*), whereas in most other plants and animals it is short range. However, molecular cloning has revealed that many organisms have both short- and long-range interspersion patterns superimposed upon one another.

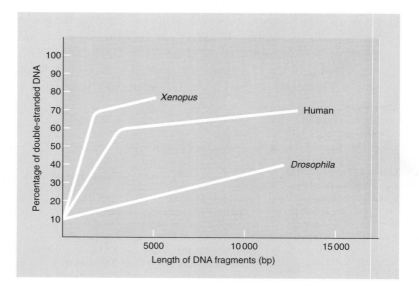

Figure 2.17 Interspersion analysis of DNA reassociation kinetics. See text for details.

The short and long interspersed repeated sequences are examples of *retroposons* (Grandbastien 1992; McDonald 1993; Bennetzen 1996). The retroviruses are the paradigm for retroposons that have the capacity to transpose because they code for reverse transcriptase and/or integrase activities. The retroposons differ from the retroviruses themselves in not passing through an independent infectious form but otherwise resemble them in the mechanism used for transposition. This group is called the *viral superfamily* of retroposons and examples are the Ty1–Ty5 elements of yeast, the *copia* and *gypsy* elements of *Drosophila* and the mammalian LINEs (long interspersed nuclear elements). These retroposons are characterized by the presence of two open-reading frames (ORFs) on one strand, one of which encodes reverse transcriptase. The viral superfamily of retroposons can be divided into two groups: those with

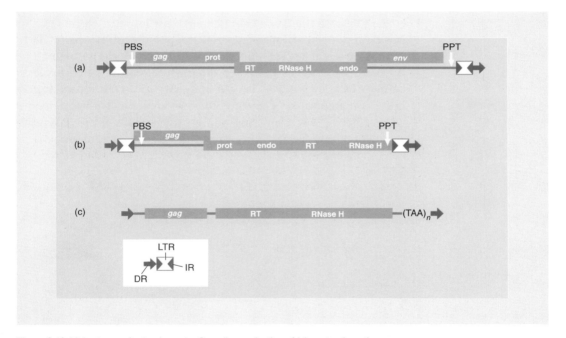

Figure 2.18 Major types of retroelements. Overall organization of (a) a retrovirus, the avian leukosis virus, (b) a retroposon, the yeast Ty1 element, and (c) a non-LTR (long terminal repeat) retroposon, the *Drosophila* I factor. Open-reading frames are depicted by green boxes. The *gag* gene encodes structural proteins of the virion core, including a nucleic-acid-binding protein; the *env* gene encodes a structural envelope protein, necessary for cell-to-cell movement; prot, protease involved in cleavage of primary translation products; RT, reverse transcriptase; RNase H, ribonuclease; endo, endonuclease necessary for integration in the host genome; LTR, long terminal repeats containing signals for initiation and termination of transcription, and bordered by short inverted repeats (IRs) typically terminating in 5′–TG. . .CA–3′; PBS, primer binding site, complementary to the 3′ end of a host tRNA, and used for synthesis of the first (–) DNA strand; PPT, polypurine tract used for synthesis of the second (+) DNA strand; DR, short direct repeats of the host target DNA, created upon insertion. (Redrawn with permission from Grandbastien 1992.)

long terminal repeats, e.g. Ty1, and those without long terminal repeats, e.g. LINEs (Fig. 2.18).

The Ty1/*copia* and Ty3/*gypsy* class retroelements have been found in all animal, plant and fungal species that have been examined but their organization and ubiquity show extensive variation. For example, the Ty1/*copia* group of retroposons are found in all plants including single-cell algae, bryophytes, gymnosperms and angiosperms. They usually are present in high copy numbers, from hundreds to millions, and as highly heterogeneous populations. This is in marked contrast to insect and fungal systems where these retroposons are present in much lower numbers (tens to hundreds) and as more homogeneous populations. There is no relationship between the total copy number and the host genome size and the copy number can vary widely between closely related species within a genus (Kumar 1996).

In *S. cerevisiae*, Ty1–Ty5 retroposons have a strong bias for sites in the genome into which they integrate (Goffeau *et al*. 1996). Over 90% of the Ty1–Ty4 elements are located within 750 bp upstream of genes transcribed by RNA polymerase III, particularly tRNA genes. The Ty5 elements are located at the telomeres or regions that have telomeric chromatin. Regions targeted by yeast retroposons are typically devoid of open-reading frames and reiterative integration can generate blocks of elements within elements (Fig. 2.19). These element landing pads provide a safe haven for elements to integrate without causing deleterious mutations.

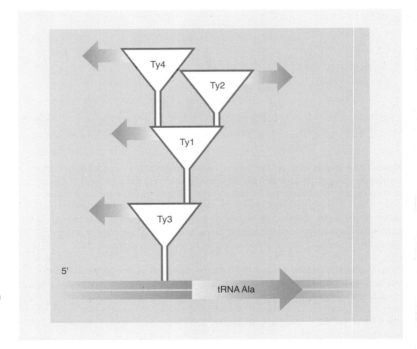

Figure 2.19 Example from *Saccharomyces cerevisiae* of retrotransposons within retrotransposons which typically are found upstream of tRNA genes. (Redrawn with permission from Voytas 1996, © American Association for the Advancement of Science.)

LINEs pervade mammalian genomes. The human genome contains about 100 000 LINEs but most are functionally inactive due to truncations, rearrangements and nonsense mutations. In the human genome about 3–4000 are full length and about 1–2% are able to transpose (Sassaman *et al.* 1997).

The fruit fly *Drosophila* appears to differ from many other organisms in having an unusually elaborate method for forming chromosome ends. It has no telomerase repeats but uses instead telomere-specific transposable elements such as *HeT-A* and *TART*. They most closely resemble LINEs (Pardue *et al.* 1996).

Another group of retroposons is the *non-viral superfamily* which does not code for proteins that have transposition functions. Rather, they have features that suggest they originated in RNA sequences. The best known examples are the mammalian SINEs (short, interspersed nuclear elements). They appear to be processed pseudogenes (see p. 24) derived from genes that encode small cytoplasmic RNA including tRNAs and 7SL RNA. The best characterized SINE family is the Alu family found in Old World primates and named for an *Alu*I restriction endonuclease site typical of the sequence. Alu units are found in nearly a million copies per haploid genome and can be found flanking genes, in introns, within satellite DNA, and in clusters with other interspersed repeated sequences. They have an observed average spacing distance of 3 kb but there are local regions of preference or exclusion (Moyzis *et al.* 1989). SINEs have been used in phylogenetic studies; for example, Shimamura *et al.* (1997) used an analysis of SINE retropositional events to confirm that whales are closely related to even-toed ungulates (cows, etc.). Within the human genome, LINEs are preferentially located in the dark G-bands of metaphase chromosomes while SINEs are preferentially found in the light G-bands.

CpG islands were originally identified as short regions of mammalian DNA which contained many sites for the restriction endonuclease *Hpa*II and for this reason originally were called *Hpa*II Tiny Fragment (HTF) islands. This island DNA is found in short regions of 1–2kb which together account for approximately 2% of the mammalian genome. It has distinctive properties when compared with the DNA in the rest of the genome. It is unmethylated, GC-rich and does not show any suppression of the dinucleotide CpG. By contrast, bulk genomic DNA has a GC content which is much lower, is methylated at CpG and the CpG dinucleotide is present at a much lower frequency than would be predicted from base composition. CpG islands have been found at the 5′ ends of all housekeeping genes and of a large proportion of genes with a tissue-restricted pattern of expression (Craig & Bickmore 1994).

CHAPTER 2
The organization and
structure of genomes

CENTROMERE STRUCTURE

The structure of centromeric DNA differs widely in different organisms. In *S. cerevisiae* it is only 170 bp long whereas in *S. pombe* it is much longer (tens of kilobases). Centromeric DNA appears to be very complex, at least in mammals where it has been extensively studied (Tyler-Smith & Willard 1993). Mammalian centromeres are made up of very large regions of repeated sequences. No unique sequence has been found in a mammalian centromere. The most extensively studied repeated-sequence families are the tandemly repeated satellite DNAs but many non-satellite centromeric repeated-sequence families also are present. The detail for the two human centromeres is shown in Fig. 2.20.

SIGNIFICANCE OF REPEAT SEQUENCES FOR GENOME MAPPING AND SEQUENCING

Although advantage can sometimes be taken of repeat sequences, generally speaking they tend to cause problems. For example, when DNA containing repeated sequences is used as a hybridization probe it will anneal to many different regions of the genome. During cloning of genome fragments recombination can occur between repeats leading to 'scrambling' of DNA sequences. During the actual DNA sequencing reactions, slippage can occur, particularly with microsatellites. Finally, during data assembly, incorrect positioning of genome fragments or sequences can occur because repeat units are incorrectly recognized as being unique. These issues are discussed in more detail in subsequent chapters.

SUMMARY OF CHROMOSOME STRUCTURES

The different structural elements that have been discussed are

Figure 2.20 Structures of the centromeres of human chromosomes 9 and 21, showing tandemly repeated satellite DNA and other centromeric repeated sequences. The sequences shown in boxes are known to be present but their precise locations have not yet been mapped. Satellite DNA is named by the size of the repeat unit except for alphoid DNA which has a 171 bp repeat. ATRS, A + T rich sequence. (Redrawn with permission from Tyler-Smith & Willard 1993.)

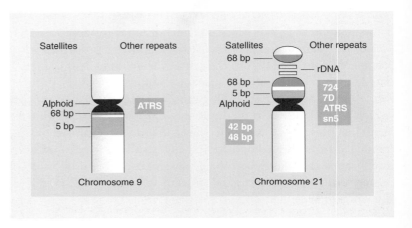

summarized in Fig. 2.21. Note that now that the genome of a eukaryote (*Saccharomyces cerevisiae*) has been completely sequenced it is possible to analyse chromosomal structure in detail (Dujon 1996). Key features elucidated to date are shown in Table 2.6.

Genome rearrangements

Hypotrichs are a large group of ciliated protozoa that cut, splice, reorder and eliminate DNA sequences to an extraordinary extent during their sexual life cycle (Prescott 1994). Similar kinds of DNA processing also occur in *Tetrohymena*. In hypotrichs there are two nuclei: a germ-line nucleus (micronucleus) and a somatic nucleus (macronucleus). Following cell mating a copy of the micronucleus gives rise to a new macronucleus and the old macronucleus is destroyed. All genes in the micronuclear genome are interrupted by non-functional sequences that are spliced out of the DNA to make functional macronuclear genes during development. Functional segments within some micronuclear genes are

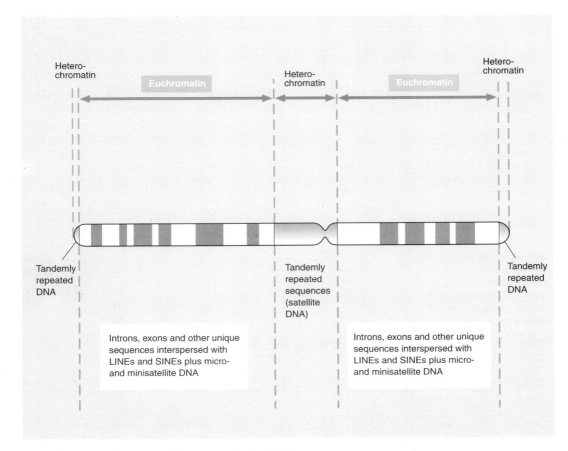

Figure 2.21 The location of repeated sequences within a typical chromosome.

43

scrambled and these segments are spliced into a different order during development to yield functional macronuclear genes. A massive elimination of micronuclear sequences yields a macronucleus with a much lower sequence complexity. Finally, during processing every macronuclear gene comes to reside in a physically separate small DNA molecule which is amplified about 1000-fold.

A different kind of programmed rearrangement occurs during antibody synthesis in man. The protein sequence of an antibody is encoded by *V*, *J* and *C* genes which are physically separated on the same chromosome. There are hundreds of different *V* genes, a limited number of *J* genes and a single *C* gene for each of the heavy and light antibody chains. For an antibody to be synthesized, a single *V* sequence has to be selected and joined to one of the *J* genes and the *C* gene for both the heavy and light chains. This complicated genetic arrangement gives the human body the opportunity to synthesize any of a vast repertoire of antibodies while minimizing the number of genes involved.

Table 2.6 The yeast genome broken down into its various identified genetic elements. (Reproduced with permission from Dujon 1996)

Genetic element	Total number	Density† (kb per element)	Average size (kb)	Total kb	Contribution to yeast genome‡ (%)
Protein-coding elements					
ORFs (total§)	6183	2.0	1.450	8965.0	68.03
Actual protein coding genes¶	5800	2.1	1.450	8410.0	63.82
Introns	233	52.0	0.500	116.0	0.88
RNA-coding elements					
rDNA repeats	120–130	–	9.000	1125.0	8.54
tRNA genes	262	46.0	0.080	21.0	0.16
Introns in tRNA genes	80	–	0.020	1.6	0.01
Other RNA species	37	326.0	0.500	18.0	0.14
Mobile elements††					
Complete Ty elements	53	227.0	5.600	297.0	2.25
Solo LTRs	363	33.0	0.330	119.0	0.90
Chromosomal elements					
ARS consensus	750	16.0	0.020	15.0	0.11
Centromeres	16	–	0.095	1.5	0.01
Subtelomeric elements (Y')††	21	–	5.800	121.0	0.92
Subtelomeric elements (X)††	31	–	0.400	12.0	0.09
Telomeric ($C_{1-3}A$) repeats	32	–	0.300	9.6	0.07
'Intergenic' regions					
Including promoters, terminators regulatory sequences, and all unidentified elements	–	–	0.500	2912.0	22.10

†Density is calculated for the whole yeast genome, except rDNA.
‡Relative contribution is calculated from the whole yeast genome, including rDNA.
§ORFs as defined in the European sequencing program.
¶Predicted number of actual protein-coding genes, assuming that 6–7% of ORFs are not actual genes.
††Actual number of such elements varies from strain to strain.

3 Subdividing the genome

Introduction

As outlined in Chapter 1, the first step in sequencing a genome is to divide the individual chromosomes in an ordered manner into smaller and smaller pieces that ultimately can be sequenced. That is, one begins by creating a genomic library. However, at some stage the different clones have to be ordered into a physical map corresponding to that found in the intact organism. The magnitude of this task depends on the average size of the cloned insert in the library: the larger the insert, the fewer clones that have to be ordered.

The number of clones required can be calculated easily. If the average size of the cloned insert is 20 kb and the genome has a size of 2.8×10^6 kb, e.g. the human genome, then the size of the genome relative to the size of the cloned insert, designated n, is $2.8 \times 10^6/20 = 1.4 \times 10^5$. The number of independent recombinants required in the library must be greater than n. This is because sampling variation will lead to the inclusion of some sequences several times and the exclusion of other sequences in a library of just n recombinants. Clarke & Carbon (1976) have derived a formula that relates the probability (P) of including any DNA sequence in a random library of N independent recombinants:

$$N = \frac{\ln(1 - P)}{\ln(1 - 1/n)}.$$

Therefore, to achieve a 95% probability ($P = 0.95$) of including any particular sequence in a random human genomic library of 20 kb fragment size:

$$N = \frac{\ln(1 - 0.95)}{\ln(1 - 1/1.4 \times 10^5)} = 4.2 \times 10^5.$$

If the probability is to be increased to 99%, then N becomes 6.5×10^5. Put a different way, a threefold coverage gives a 95% probability of including any sequencing and a fivefold coverage a 99% probability. These calculations assume equal representation of sequences, but this is not true in practice.

Fragmentation of DNA with restriction enzymes

Type II restriction endonucleases have target sites which are 4–8 bp in length. If all bases are equally frequent in a DNA molecule then we would expect a tetranucleotide to occur on average every 4^4 (i.e. 256) nucleotide pairs in a long random DNA sequence. Similarly, a hexanucleotide would occur every 4^6 (i.e. 4096) bp and an octanucleotide every 4^8 (i.e. 65 536) bp. Table 3.1 shows the *expected* number of fragments that would be produced when different genomes are digested completely with different restriction endonucleases. In practice, the actual number of fragments is quite different because the distribution of nucleotides is non-random and most organisms do not have an equal number of the four bases; for example, human DNA has overall only 40% G + C and the frequency of the dinucleotide CpG is only 20% of that expected. In addition, many cytosine residues are methylated and this can prevent restriction endonuclease digestion. Thus the enzyme *Not*I, which recognizes the sequence GCGGCCGC, cuts human DNA into fragments of average size 1000–1500 kb rather than the 65 kb expected. Similarly, the *E. coli* genome is cut by *Not*I into only 20 fragments, not the 72 expected from Table 3.1 (Smith *et al.* 1987). Again, the *Schizosaccharomyces pombe* genome, which is slightly smaller than that of *Saccharomyces cerevisiae*, is cut into only 14 fragments (Fan *et al.* 1989). In addition to *Not*I there are a number of other restriction endonucleases with 8-bp recognition sequences (Table 3.2). In genomes rich in G + C the tetranucleotide CTAG is particularly rare (McClelland *et al.* 1987), such that the enzymes *Spe*I (ACTAGT), *Xba*I (TCTAGA), *Nhe*I (GCTAGC) and *Bln*I (CCTAGG) can produce a more limited number of DNA fragments than might at first be expected.

Gelfand and Koonin (1997) have analysed those bacterial genomes that have been completely sequenced and have found that short palindromic sequences, like restriction endonuclease recognition sites, are deficient at a statistically significant level. They suggest that in the course of evolution bacterial DNA has been exposed to a wide spectrum of restriction enzymes, probably as a

Table 3.1 Expected number of DNA fragments produced from different genomes by restriction endonucleases with tetra-, hexa- and octanucleotide recognition sequences. The expected number assumes that the DNA has a 50% G + C content and there is a totally random distribution of the four bases along any one strand of DNA

Organism	Haploid genome size (n)	Expected number of fragments		
		4-Cutter (n/256)	6-Cutter (n/4096)	8-Cutter (n/65 536)
Escherichia coli	4.70×10^6 bp	18 359	1147	72
Saccharomyces cerevisiae	1.35×10^7 bp	52 734	3296	206
Drosophila melanogaster	1.80×10^8 bp	703 125	43 945	2746
Homo sapiens	2.80×10^9 bp	1093 750	683 593	42 274

Table 3.2 Restriction endonucleases with 8-bp recognition sequences

Endonuclease	Recognition sequence
NotI	GCGGCCGC
SfiI	GGCCnnnnGGCC
SwaI	ATTTAAAT
PacI	TTAATTAA
PmeI	GTTTAAAC
SgrAI	CACCGGTG
Sse83871	CCTGCAGG
SrfI	GCCCGGGC
SgfI	GCGATCGC
FseI	GGCCGGCC
AscI	GGCGCGCC

result of latent transfer mediated by mobile genetic elements.

The average size of DNA fragment produced by digestion with restriction enzymes with 4- and 6-base recognition sequences is too small to be of much use for preparing gene libraries except in special circumstances (see below). Even enzymes with 8-base recognition sequences may not be of particular value because, although the average size of fragment should be 65.5 kb, in the case of *S. pombe*, in practice the fragments range in size from 4.5 kb to 3.5 Mb (Fan *et al*. 1989). If more uniform-sized fragments are required it is usual to partially digest the target DNA with an enzyme with a 4-base recognition sequence. The partial digest then can be fractionated (see next section) to separate out fragments of the desired size. Since the DNA is randomly fragmented there will be no exclusion of any sequence. Furthermore, clones will overlap one another (Fig. 3.1) and this is particularly important when trying to order different clones into a map (see Chapter 4).

Some introns encode endonucleases that are site-specific and have 18–30 bp recognition sequences (Dujon *et al*. 1989). These endonucleases can be used to produce a very limited number of fragments, some or all of which are produced by cleavage within related genes. For example, the intron-encoded endonuclease

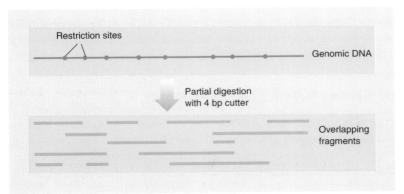

Figure 3.1 The generation of overlapping fragments of genomic DNA following partial digestion with a restriction enzyme with a tetranucleotide recognition sequence.

I-*Cen*I from the chloroplast large rRNA gene of *Chlamydomonas eugametos* cuts the chromosomes of *E. coli* and *Salmonella* seven times, once within each of the seven *rrn* operons (Liu *et al.* 1993). The I-*Sce*I intron encoded endonuclease from *Saccharomyces cerevisiae* has an 18 bp recognition site and the VDE endonuclease from the same organism has a 30 bp recognition site. Assuming a random organization of sequences, the frequency of occurrence of an 18 bp site is one per 7×10^{10} bp (4^{18}). Thus there are probably few or no endogenous sites even within very complex genomes (Jasin 1996).

New sites for rare-cutting endonucleases can be introduced into genomes to facilitate gene mapping. For example, in bacteria a number of temperate phages and transposons contain single *Not*I sites and their insertion into the chromosome will create one or more new sites depending on the number of integration events (Smith *et al.* 1987; Le Bourgeois *et al.* 1992). Similarly, an integrative vector was used to generate sites in yeast for the intron-encoded endonuclease I-*Sce*I (Thierry & Dujon 1992). A variation of this technique can be used to reduce the number of cleavage sites. A novel *E. coli lac* operator containing a site for *Hae*II (Pu GCGC Py) was introduced into the genomes of *E. coli* and yeast. Addition of the *lac* repressor protein protects this site from methylation by the M.*Hha* methyltransferase which recognizes the sequence GCGC found in all *Hae*II recognition sites. Consequently, a single site remains that is susceptible to cleavage by *Hae*II, the so-called *Achilles heel* site (Koob & Szbalski 1990).

The RecA-assisted restriction endonuclease (RARE) technique for reducing the number of cleavage sites involves the RecA-mediated formation of a triplex structure between an oligonucleotide and a chosen locus on the chromosome containing the site of a given restriction enzyme (Koob & Szbalski 1990; Ferrin & Camerini-Otero 1991). After methylation of the genome by the corresponding methyltransferase and removal of the RecA protein, only the protected site is cleaved by the endonuclease. Two such targeted cuts permit the excision of a unique DNA fragment from the genome (Szybalski 1997) and this methodology has been used to facilitate optical mapping (see p. 89). Although these methods for reducing the number of restriction sites potentially are very powerful, they are technically difficult and are not widely used.

Separating large fragments of DNA

As a DNA molecule is digested with a restriction endonuclease it is cut into smaller and smaller pieces. As such, the number of smaller molecules will always exceed the number of larger molecules. Small fragments ligate more efficiently and clones with small inserts transform with a higher efficiency. If there is no size selection then in a random cloning exercise the recombinants will have a preponderance of small inserts. Some vectors have an automatic size selection, e.g. the λ phage (5–25 kb) and cosmid (35–45 kb) vectors, but most do not. Thus it is common practice after DNA digestion to isolate DNA fragments of the desired size.

The most widely used method of separating large DNA molecules is by electrophoresis in agarose. An agarose gel is a complex network of polymeric molecules whose average pore size is dependent on the buffer composition and the type and composition of agarose used. Upon electrophoresis, DNA molecules display elastic behaviour by stretching in the direction of the applied field and then contracting into balls. The larger the pore size of the gel, the greater the ball of DNA that can pass through and hence the larger the molecules that can be separated, i.e. the gel acts as a sieve. Once the globular volume of the DNA molecule exceeds the pore size, the DNA molecule can only pass through by reptation (i.e. a snake-like movement). This leads to size-independent mobility which occurs with conventional agarose gel electrophoresis above 20–30 kb. To achieve reproducible separations of DNA fragments larger than this it is necessary to use pulsed electrical fields.

Pulsed field gel electrophoresis (PFGE) was developed by Schwartz & Cantor (1984). The technique employs alternately pulsed perpendicularly oriented electrical fields (Fig. 3.2). The duration of the applied electrical pulses is varied from 1 s to 90 s permitting separation of DNAs with sizes from 30 kb to 2 Mb. The mechanism by which this occurs is believed to be as follows. When a large DNA molecule enters a gel in response to an electrical field it must elongate parallel to the field. This field then is cut off and a new field applied at right angles to the long axis of the DNA. The molecule is unable to move until it re-orientates in a new direction and the time required for this will depend on molecular weight. Repeating the cycle results in each DNA molecule having a characteristic net mobility along the diagonal of the gel. The use of non-uniform electrical fields is critical in achieving high resolution. The reason for this is that the leading edge of a DNA band is exposed to a weaker electrical field than the trailing edge and thus the band is subject to constant compression. The usefulness of this method has been demonstrated by the separation of the different fragments of DNA produced by *Not*I digestion of *E. coli* and *S. pombe* DNA (Smith *et al.* 1987; Fan *et al.* 1989) and the separation of intact *S. cerevisiae* chromosomes (Schwartz & Cantor 1984).

A major disadvantage of PFGE, as originally described, is that the DNA samples do not run in straight lines, making analysis by Southern blotting, etc., difficult. This problem has been overcome by the development of improved methods for alternating electrical fields including orthogonal-field-alteration gel electrophoresis (OFAGE), field-inversion gel electrophoresis (FIGE), transversely

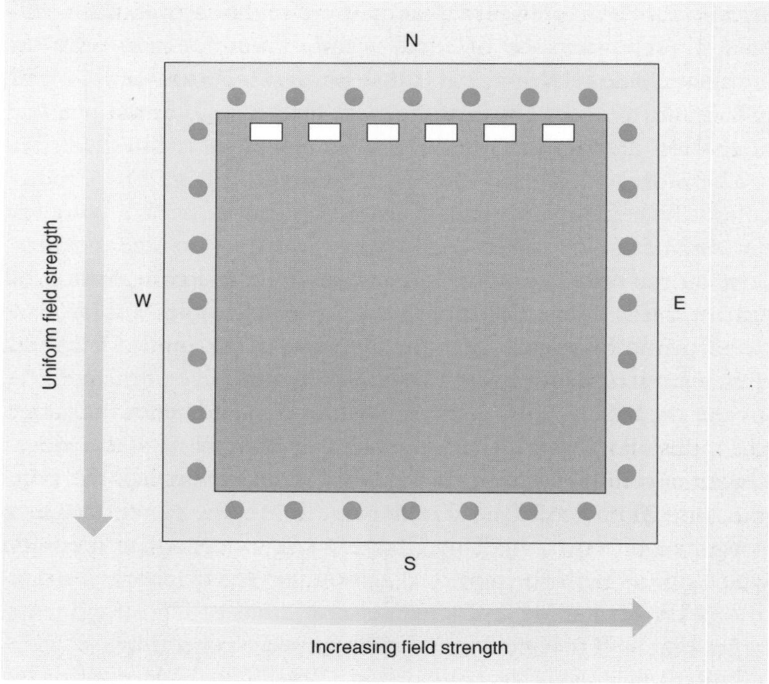

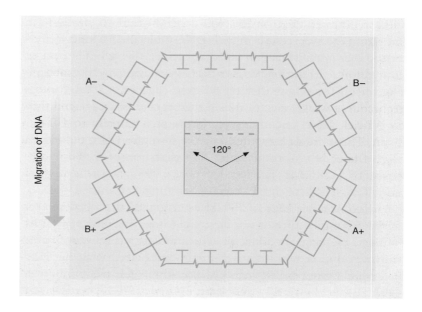

Figure 3.2 Instrumentation for pulsed field gradient electrophoresis. The platinum electrodes are shown as green dots and the gel slots are positioned along one edge of the gel matrix which is shown in grey. The letters N, S, E and W refer to the electrical field orientation. (Adapted from Schwartz & Cantor 1984.)

alternating-field electrophoresis (TAFE) and contour-clamped homogeneous electrical field electrophoresis (CHEF) (Cantor *et al*. 1988). CHEF, which was first described by Chu *et al*. (1986), is the most commonly used method for genomes of all types. It uses a hexagonal array of fixed electrodes (Fig. 3.3) and this creates a homogeneous electrical field resulting in enhanced resolution of DNA fragments. The DNA tracks run in straight lines and are

Figure 3.3 Schematic representation of CHEF (contour-clamped homogeneous electrical field) pulsed field gel electrophoresis.

comparable across the gel, thus making calibration with markers and the interpretation of Southern blots simpler than with the original PFGE. With these developments the reproducible size resolution has been increased to 10 Mb which is greater in size than most prokaryotic genomes.

Isolation of chromosomes

The task of lining up the different clones that make up a gene library can be simplified if the individual chromosomes are separated prior to digestion. PFGE permits the separation of yeast chromosomes (Fan *et al*. 1989), the largest of which is 3–5 Mb. However, the chromosomes of higher eukaryotes are much larger than this. For example, *Drosophila* chromosome 2 is 67 Mb in length and human chromosomes vary in size from 50 Mb (chromosome 21) to 263 Mb (chromosome 1). Electrophoresis cannot separate molecules this large. Instead, fluorescence-activated cell sorting (FACS), also known as flow karyotyping, has to be used (Davies *et al*. 1981).

In FACS, chromosome preparations are stained with a DNA binding dye which can fluoresce in a laser beam. The amount of fluorescence exhibited by a given chromosome is proportional to the amount of dye bound. This, in turn, is proportional to the amount of the DNA and hence the size of the chromosome. Chromosomes can therefore be fractionated by size in a FACS machine (Fig. 3.4). A stream of droplets containing single stained chromosomes is passed through a laser beam at the rate of 2000 chromosomes per second and the fluorescence from each is monitored by a photomultiplier. When the fluorescence intensity indicates that the chromosome illuminated by the laser is the one desired, the charging collar puts an electrical charge on the droplet. When droplets containing the desired chromosome pass between charged deflection plates, they are deflected into a collection vessel. Uncharged droplets lacking the desired chromosome pass into a waste collection vessel. In a variation of the above technique, two fluorescent dyes are used: one binds preferentially to AT-rich DNA and the other binds preferentially to GC-rich DNA. The stained chromosomes pass through a point on which a pair of laser beams are focused, one beam to excite the fluorescence of each dye. Each chromosome type has characteristic numbers of AT and GC base pairs so chromosomes can be identified by a combination of the total fluorescence and the ratio of the intensities of the fluorescence emissions from the two dyes.

The kind of separation that can be achieved with human chromosomes is shown in Fig. 3.5. Note that some of the chromosomes cannot be separated in the FACS because they are of similar size and AT/GC ratio. However, separation can

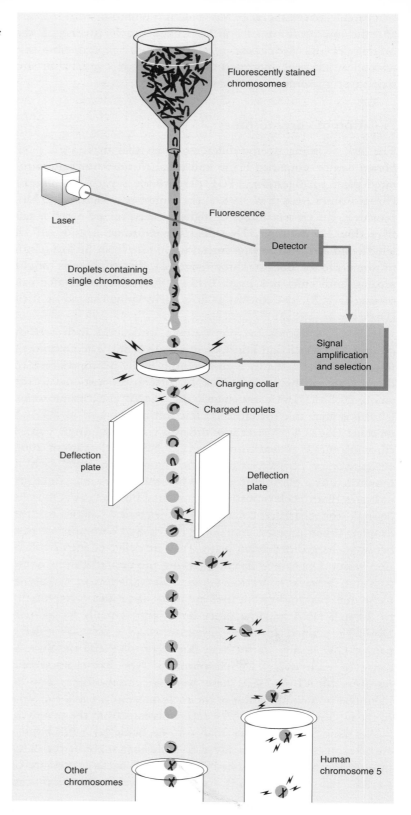

Fluorescently stained
chromosomes

Laser

Fluorescence

Detector

Droplets containing
single chromosomes

Signal
amplification
and selection

Charging collar

Charged droplets

Deflection
plate

Deflection
plate

Other
chromosomes

Human
chromosome 5

Figure 3.4 Schematic
representation of single
chromosome separation by
FACS. See text for details.
(Redrawn with permission
from Dogget 1992, courtesy of
University Science Books.)

beachieved if *somatic cell hybrids* are used as the source of DNA (D'Eustachio & Ruddle 1983). These hybrids contain a full complement of chromosomes of one species but only one or a limited number of chromosomes from the second. They are formed by fusing the cells of the two different species and applying conditions that select against the two donor cells. One of the parents may be sensitive to a particular drug and the other might be a mutant requiring special conditions for growth, e.g. thymidine kinase negative cells do not grow in HAT medium. In the presence of the drug and HAT medium, only hybrid cells containing a functional thymidine kinase gene can grow. If hybrid cells are grown under non-selective conditions after the initial selection, chromosomes from one of the parents tend to be lost more or less at random. In the case of human/rodent fusions, which are the most common, the human chromosomes are preferentially lost. Eventually only one or a few chromosomes from one parent remain. In this way rodent cell lines containing one or two human chromosomes have been constructed and these can be used to isolate the individual human chromosomes by FACS.

Lee *et al.* (1994) have shown that chromosome sorting from hybrid cells can be simplified if rodent cells are replaced with cells from the Indian Muntjac deer. Cells of the Muntjac deer have only a very small number of giant chromosomes: two autosomes plus X and Y. Thus donor chromosomes in hybrid cells are easily

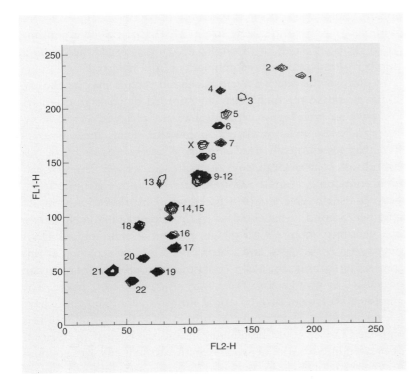

Figure 3.5 Separation of different human chromosomes by FACS using two fluorescent dyes, Hoechst 33258 (FL1-H) and chromomycin A3 (FL2-H). (Kindly supplied by Dr D. Davies, Imperial Cancer Research Fund.)

separated from host chromosomes. Not only were highly purified human chromosomes obtained by flow cytometry but they were obtained at over 90% purity by rate zonal centrifugation in sucrose gradients.

Chromosome microdissection

Particular subchromosomal regions, such as a chromosomal band (see Fig. 2.21), can be obtained by microdissection of metaphase chromosomes. A micromanipulator with very fine needles is used to cut out the desired band from individual chromosomes. When sufficient material has been collected the DNA is cloned. The technique originally was developed for use with *Drosophila* and mouse DNA and at least 100 chromosomes were needed for dissection. Ludecke *et al.* (1989) have simplified this technique. Only a few chromosomes are dissected and after cloning into a vector this DNA is amplified by means of PCR.

Vectors for cloning DNA

Once the genome has been fragmented it is essential to propagate each fragment to enable it to be mapped and ultimately sequenced. Although many different types of vector have been developed (see Old & Primrose 1994, for review) only a few are of use for large-scale genome sequencing projects. The early work on construction of gene libraries made use of bacteriophage λ vectors. With such vectors the largest size of DNA that can be accommodated is about 25 kb. If *in vitro* packaging is used, then a large number of independent recombinants can be selected.

In place of the phage λ vectors, cosmid vectors may be chosen. These also have the high efficiency afforded by packaging *in vitro* and have an even higher capacity than any phage λ vector. However, there are two drawbacks in practice. First, most workers find that screening libraries of phage λ recombinants by plaque hybridization gives cleaner results than screening libraries of bacteria containing cosmid recombinants by colony hybridization. Plaques usually give less of a background hybridization than do colonies. Second, it may be desirable to retain and store an amplified genomic library. With phage, the initial recombinant DNA population is packaged and plated out. It can be screened at this stage. Alternatively, the plates containing recombinant plaques can be washed to give an *amplified* library of recombinant phage. The amplified library can then be stored almost indefinitely; phage have a long shelf-life. The amplification is so great that samples of this amplified library could be plated out and screened with different probes on hundreds of occasions. With bacterial colonies containing cosmids it is also possible to store an amplified library

(Hanahan & Meselson 1980), but bacterial populations cannot be stored as readily as phage populations. There is often an unacceptable loss of viability when the bacteria are stored.

A word of caution is necessary when considering the use of any amplified library. There is the possibility of *distortion*. Not all recombinants in a population will propagate equally well, e.g. variations in target DNA size or sequence may affect replication of a recombinant phage, plasmid or cosmid. Therefore, when a library is put through an amplification step particular recombinants may be increased in frequency, decreased in frequency, or lost altogether. Factors which affect sub-clone representation include the nature and complexity of repeat sequences, length of the repeat region and insert orientation (Chissoe *et al.* 1997). Notable differences in sub-clone representation also can occur between related vectors. Development of modern vectors and cloning strategies has simplified library construction to the point where many workers now prefer to create a new library for each screening, rather than risk using a previously amplified one.

Genomic DNA libraries in phage λ vectors are expected to contain most of the sequences of the genome from which they have been derived. However, deletions can occur, particularly with DNA containing repeated sequences. As noted in Chapter 2, repeated sequences are widespread in the DNA from higher eukaryotes. The deletion of these repeated sequences is not prevented by the use of recombination-deficient strains.

Yeast artificial chromosomes (YACs)

The upper size limit of 35–45 kb for cloning in a cosmid means that it would take 4500 clones to cover the *D. melanogaster* genome and 70 000 clones to cover the human genome. What is required is a vector that permits a larger insert size and YACs make this possible. It will be recalled from Chapter 2 that the minimum structural elements for a linear chromosome are an origin of replication (*ars*), telomeres and a centromere. Murray and Szostak (1983) combined an *ars* and a centromere from yeast with telomeres from *Tetrahymena* to generate a linear molecule that behaved as a chromosome in yeast. When the requirements for normal replication and segregation were studied it was found that the length of the YAC was important. When the YAC was less than 20 kb in size, centromere function was impaired. However, much larger YACs segregated normally. Burke *et al.* (1987) made use of this fact in developing a vector (Fig. 3.6) for cloning large DNA molecules. They showed that YACs could be used to generate whole libraries from the genomes of higher organisms with insert sizes at least tenfold larger than those that can be accommodated by bacteriophage λ and cosmid vectors.

A major drawback with YACs is that although they are capable of replicating large fragments of DNA, manipulating such large DNA fragments in the liquid phase prior to transformation, and keeping them intact, is very difficult. Thus many of the early YAC libraries had average insert sizes of only 50–100 kb. By removing small DNA fragments by PFGE fractionation prior to cloning, Anand *et al.* (1990) were able to increase the average insert size to 350 kb. By including polyamines to prevent DNA degradation, Larin *et al.* (1991) were able to construct YAC libraries from mouse and human DNA with average insert sizes of 700 and 620 kb, respectively. More recently, Bellanné-Chantelot *et al.* (1992) constructed a human library with an average insert size of 810 kb and with some inserts as large as 1800 kb.

There are a number of operational problems associated with the

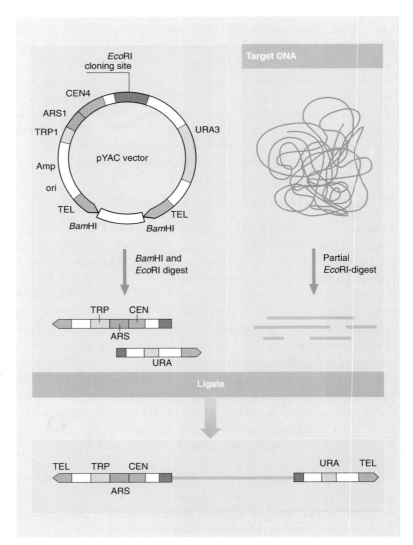

Figure 3.6 Construction of a YAC containing large pieces of cloned DNA. Key regions of the pYAC vector are as follows: TEL, yeast telomeres; ARS1, autonomously replicating sequence; CEN4, centromere from chromosome 4; URA3 and TRP1, yeast marker genes; Amp, ampicillin-resistance determinant of pBR322; ori, origin of replication of pBR322.

use of YACs (Kouprina *et al*. 1994; Monaco & Larin 1994). The first of these is that it is estimated that 10–60% of clones in existing libraries represent chimaeric DNA sequences: that is, sequences from different regions of the genome cloned into a single YAC. Chimaeras may arise by co-ligation of DNA inserts *in vitro* prior to yeast transformation, or by recombination between two DNA molecules that were introduced into the same yeast cell. It is possible to detect chimaeras by *in situ* hybridization of the YAC to metaphase chromosomes: hybridization to two or more chromosomes or to geographically disparate regions of the same chromosome is indicative of a chimaera.

A second problem with YACs is that many clones are unstable and tend to delete internal regions from their inserts. Using a model system, Kouprina *et al*. (1994) were able to show that deletions can be generated both during the transformation process and during mitotic growth of transformants and that the size of the deletions varied from 20 to 260 kb. Ling *et al*. (1993) showed that the frequency of deletion formation could be reduced by use of a strain rendered recombination-deficient due to a *rad52* mutation. However such strains grow more slowly and transform less efficiently than RAD$^+$ strains and therefore are not ideal hosts for YAC library construction. Le & Dobson (1997) have shown that the *rad54-3* allele significantly stabilizes YAC clones containing human satellite DNA sequences. Strains carrying this allele can undergo meiosis and have growth and transformation rates comparable with wild-type strains. Heale *et al*. (1994) have shovn that chimaera formation results from the yeast's mitotic recombination system which is stimulated by the spheroplasting step of the standard YAC transformation system. Transformation of intact yeast cells is much less recombinagenic. An additional limitation on the use of YACs is the high rate of loss of some YACs during mitotic growth.

The third major problem with YAC clones is that the 15 Mb yeast host chromosome background cannot be separated from the YACs by simple methods. Nor is the yield of DNA very high. Unlike plasmid vectors in bacteria, YACs have a structure very similar to natural yeast chromosomes. Thus purifying YAC from the yeast chromosomes usually requires separation by PFGE. Alternatively, the entire yeast genome is subcloned in bacteriophage or cosmid vectors followed by identification of those clones carrying the original YAC insert.

BACs and PACs as alternatives to YACs

Phage P1 is a temperate bacteriophage which has been extensively used for genetic analysis of *E. coli* because it can mediate generalized transduction. Sternberg and co-workers have developed a P1 vector system which has a capacity for DNA fragments as large as 100 kb (Sternberg 1990; Pierce *et al.* 1992). Thus the capacity is about twice that of cosmid clones but less than that of YAC clones. The P1 vector contains a packaging site (*pac*) which is necessary for *in vitro* packaging of recombinant molecules into phage particles. Vectors contain two *lox*P sites. These are the sites recognized by the phage recombinase, the product of the phage *cre* gene, and which lead to circularization of the packaged DNA after it has been injected into an *E. coli* host expressing the recombinase (Fig. 3.7). Clones are maintained in *E. coli* as low copy number plasmids by selection for a vector kanamycin-resistance marker. A high copy number can be induced by exploitation of the P1 lytic

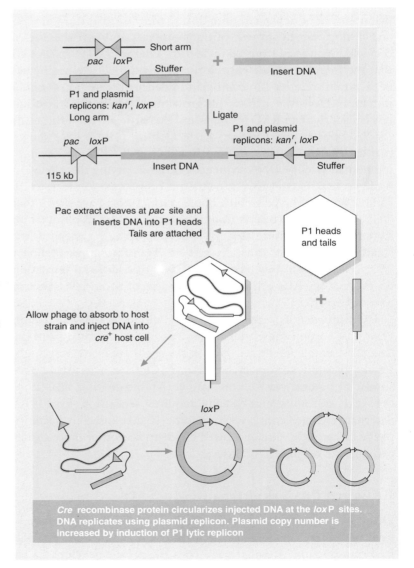

Figure 3.7 The phage P1 vector system. The P1 vector Ad10 (Sternberg 1990) is digested to generate short and long vector arms. These are dephosphorylated to prevent self-ligation. Size-selected insert DNA (85–100 kb) is ligated with vector arms, ready for a two-stage processing by packaging extracts. First, the recombinant DNA is cleaved at the *pac* site by pacase in the packaging extract. Then the pacase works in concert with head/tail extract to insert DNA into phage heads, *pac* site first, cleaving off a headful of DNA at 115 kb. Heads and tails then unite. The resulting phage particle can inject recombinant DNA into host *E. coli*. The host is *cre*⁺. The *cre* recombinase acts on *lox*P sites to produce a circular plasmid. The plasmid is maintained at low copy number, but can be amplified by inducing the P1 lytic operon.

replicon (Sternberg 1990). This P1 system has been used to construct genomic libraries of mouse, human, fission yeast and *Drosophila* DNA (Hoheisel *et al.* 1993; Hartl *et al.* 1994). Estimates of chimaerism in such libraries are well below the estimates for YAC libraries. For example, Hartl *et al.* (1994) found chimaeric clones in less than 2% of more than 3000 *Drosophila* P1 clones and Sternberg (1994) reported less than 5% in more than 1000 human and mouse P1 clones.

Shizuya *et al.* (1992) have developed a bacterial cloning system for mapping and analysis of complex genomes. This BAC system (*b*acterial *a*rtificial *c*hromosome) is based on the single-copy sex factor F of *E. coli*. This vector (Fig. 3.8) includes the λ *cos*N and P1 *lox*P sites, two cloning sites (*Hind*III and *Bam*HI) and several G + C restriction enzyme sites (e.g. *Sfi*I, *Not*I, etc.) for potential excision of the inserts. The cloning site also is flanked by T7 and SP6 promoters for generating RNA probes. This BAC can be transformed into *E. coli* very efficiently, thus avoiding the packaging extracts that are required with the P1 system. The BAC is capable of maintaining human and plant genomic fragments of greater than 300 kb for over 100 generations with a high degree of stability (Woo *et al.* 1994) and has been used to construct a rice nuclear genome library with an average insert size of 125 kb (Wang *et al.* 1995a). There are, however, two disadvantages to the

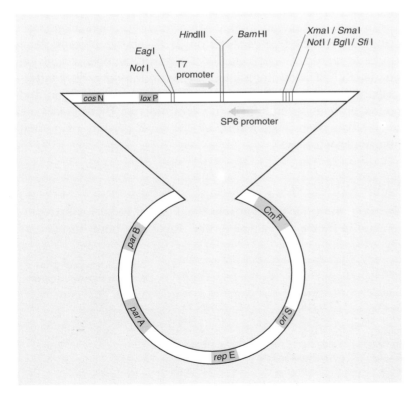

Figure 3.8 Structure of a BAC vector derived from a mini-F plasmid. The *ori*S and *rep*E genes mediate the unidirectional replication of the F factor, while *par*A and *par*B maintain the copy number at a level of one or two per genome. Cm^R is a chloramphenicol-resistance marker. *cos*N and *lox*P are the cleavage sites for λ terminase and P1 *cre* protein, respectively. *Hind*III and *Bam*HI are unique cleavage sites for inserting foreign DNA. (Adapted from Shizuya *et al.* 1992.)

use of BACs. First, there currently is no method of positively selecting clones with foreign DNA inserts. Second, because BACs cannot be amplified it is difficult to isolate large amounts of DNA.

More recently, Ioannou *et al.* (1994) have developed a P1-derived artificial chromosome, or PAC, by combining features of both the P1 and F-factor systems. The PAC vector is able to handle inserts in the 100–300 kb range and, as yet, no chimaeras or clone instability have been detected.

Human artificial chromosomes: HAECs and HACs

One problem with YACs, BACs and PACs is that they are not maintained as plasmids or minichromosomes outside their 'normal' host, i.e. yeast or *E. coli*. Rather, in foreign cells the vector DNA plus insert can be maintained only by integrating into the host chromosome. This is not ideal for studying the function of the cloned sequence, particularly if it is a functional unit many hundreds of kilobases in length. Furthermore, integration can result in damage to chromosomal genes and the integrated DNA may be rearranged, deleted or disrupted. To try and overcome this problem a number of groups have attempted to construct human artificial chromosomes (HACs). Sun *et al.* (1994) developed an episomal (HAEC) system based on the replication origin of the herpes Epstein–Barr virus. HAECs are stably maintained in human cells as circular minichromosomes and have been used to carry inserts as large as 330 kb. More recently, Harrington *et al.* (1997) constructed first-generation HACs by combining long synthetic arrays of alpha satellite DNA with telomeric and genomic DNA. The resulting linear microchromosomes were mitotically and cytogenetically stable in the absence of selection and bound proteins specific for active centromeres. These HACs are 6–10 Mb in size, i.e. one-fifth to one-tenth the size of normal human chromosomes.

Choice of vector

The maximum size of insert that the different vectors will accommodate is shown in Table 3.3, and Table 3.4 shows the size of

Vector	Host	Insert size
λ phage	*E. coli*	5–25 kb
λ cosmids	*E. coli*	35–45 kb
P1 phage	*E. coli*	70–100 kb
PACs	*E. coli*	100–300 kb
BACs	*E. coli*	≤ 300 kb
HAECs	Mammalian cells	≤ 330 kb
YACs	*S. cerevisiae*	200–2000 kb

Table 3.3 Maximum DNA insert possible with different cloning vectors

different genomes relative to the size of insert with the various vectors. However, as noted above, the size of insert is not the only feature of importance. The absence of chimaeras and deletions is even more important. In practice, some 50% of YACs show structural instability of inserts or are chimaeras in which two or more DNA fragments have become incorporated into one clone. These defective YACs are unsuitable for use as mapping and sequencing reagents and a great deal of effort is required to identify them. Cosmid inserts sometimes contain the same abberations and the greatest problem with them arises when the DNA being cloned contains tandem arrays of repeated sequences. The problem is particularly acute when the tandem array is several times larger than the allowable size of a cosmid insert. Potential advantages of the BAC and PAC systems over YACs include lower levels of chimaerism, ease of library generation and ease of manipulation and isolation of insert DNA. BAC clones seem to represent human DNA far more faithfully than their YAC or cosmid counterparts and appear to be excellent substrates for shotgun sequence analysis resulting in accurate contiguous sequence data (Venter *et al.* 1996).

Avoiding the need to isolate individual chromosomes

The construction and identification of YAC clones from specific chromosomes serves as an important step in the assembly of overlapping clone libraries of these chromosomes. Earlier (p. 51) it was described how individual chromosomes could be isolated by flow cytometry prior to cloning. However, the small amount of material generated by flow-sorting makes library construction with very large inserts extremely difficult. This leads to an increase in the number of clones required to achieve a given coverage, which partly offsets the advantage of having a chromosome-specific library.

An alternative approach is to make a YAC library from a somatic cell hybrid (see p. 53), e.g. a monosomic human–rodent hybrid. Once the library is constructed those clones that carry the human DNA are distinguished from those containing rodent DNA by hybridization with a probe specific for the former, e.g. human

Organism	Haploid genome size	Size of genome relative to size of cloned DNA (*n*)			
		Cosmid (40 kb)	P1 (85 kb)	BAC/PAC (250 kb)	YAC (1000 kb)
E. coli	4.70×10^6 bp	118	55	19	5
S. cerevisiae	1.35×10^7 bp	338	159	54	14
D. melanogaster	1.80×10^8 bp	4500	2118	720	180
H. sapiens	2.80×10^9 bp	70 000	32 941	11 200	2800

Table 3.4 Size of different genomes relative to the average DNA insert that is obtained with different cloning vectors

DNA repeat sequences (Abidi *et al.* 1990; Wada *et al.* 1990). The problem with this approach is that to isolate all 24 human chromosomes (22 autosomes plus two sex chromosomes) would require the construction of 24 individual libraries from a full panel of human monosomic hybrid cell lines. The only justification for such a daunting task is that YAC libraries derived from rodent–human cell lines *might* contain a lower frequency of chimaeric YACs than those derived from a single species. Another problem is contamination and biases in the initial flow-sorted libraries (Hudson *et al.* 1994).

A less labour-intensive approach than the one described above is to generate a complete genomic library and then divide it into chromosomal subsets. Division is achieved by screening all the clones with chromosome-specific DNA probes. The feasibility of this approach has been proven with human chromosome 21. Ross *et al.* (1992) developed their chromosome-21-specific probes from a series of phage libraries constructed from flow-sorted chromosomes. Inserts were excised from vector sequences and recloned using a plasmid vector. The plasmid clones from each chromosome were then pooled together. The purpose of this recloning procedure was to increase the ratio of insert to vector sequences. Chumakov *et al.* (1992a) developed their probes from somatic cell hybrid lines containing individual human chromosomes. Human chromosomal DNA lying between *Alu* repeats was amplified by PCR using as a primer a consensus *Alu* repeat sequence. In this way a family of chromosome-21-specific sequences was obtained free of rodent DNA. Yet a third set of chromosome-21-specific probes was developed by Chumakov *et al.* (1992b) from the known sequence of chromosome 21 genes deposited in nucleic acid databases.

4 Assembling a physical map of the genome

Introduction

After a genome has been fragmented and the fragments cloned to generate a genomic library, it is necessary to assemble the cloned fragments in the same linear order as found in the chromosomes from which they were derived. Positioning cloned DNA fragments is analogous to completing one edge of a jigsaw puzzle but, rather than looking for interlocking pieces, detectable overlaps between clones are sought, i.e. clones with a unique stretch of DNA in common. Because the number of fragments is so large a rapid method for detecting overlaps between pairs of clones is needed. If each clone could be sequenced, overlaps could be identified unambiguously, provided the overlapping region is not a sequence that repeats itself elsewhere in the genome. However, the current state of sequencing technology is such that this approach is totally impractical; for example, using a cosmid library of the human genome in which the average insert size is 40 kb, approximately 10 000 sequencing gels would be required.

One method of linking cloned fragments is *chromosome walking* (Bender *et al*. 1983) which was originally developed for the isolation of gene sequences whose function is unknown but whose genetic location is known. The principle of this method is shown in Fig. 4.1. For the purposes of map generation a single cloned fragment is selected. This is used as a probe to detect other clones in the library with which it will hybridize and which represent clones overlapping with it. The overlap can be to the right or the left. This single walking step is repeated many times and can occur in both directions along the chromosome. A potential problem with chromosome walking is created by the existence of repeated sequences. If the clone used as the probe contains a sequence repeated elsewhere in the genome, it will hybridize to non-contiguous fragments. For this reason the probe used for stepping from one genomic clone to the next must be a unique sequence clone, or a sub-clone which has been shown to contain only a unique sequence. Clearly, if chromosome walking is to be em-

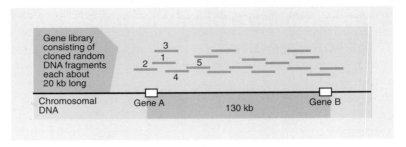

Figure 4.1 Chromosome walking. It is desired to clone DNA sequences of gene B, which has been identified genetically but for which no probe is available. Sequences of a nearby gene A are available in fragment 1. Alternatively, a sequence close to gene B could be identified by *in situ* hybridization to *Drosophila* polytene chromosomes. In a large, random genomic DNA library many overlapping cloned fragments are present. Clone 1 sequences can be used as a probe to identify overlapping clones 2, 3 and 4. Clone 4 can, in turn, be used as a probe to identify clone 5, and so on. It is, therefore, possible to walk along the chromosome until gene B is reached.

ployed it makes sense to use very large fragments of DNA as this minimizes the number of steps.

Although inherently attractive, chromosome walking is too laborious and time consuming to be of value for mapping of most genomes. Reference to Table 3.4 shows that with cosmid clones a minimum of 4500 individual 'walks' would be necessary with the *Drosophila* genome and 70 000 with the human genome. In practice, many more 'walks' would be necessary since a representative library would contain a much larger number of clones (see p. 45). Three general methodologies for mapping have been developed as alternatives to chromosome walking: restriction enzyme fingerprinting, marker sequences and hybridization assays. In each case the objective is to create a landmark map of the genome under study with markers dispersed at regular intervals throughout the map. In restriction enzyme fingerprints the markers are restriction sites; with the other methods they are unique sequences detected by the PCR or by hybridization. Each of the methods has its limitations and the current trend is to integrate the maps produced by the different methods. In some cases the generation of a map requires the use of all three methods, as was found by Chumakov *et al.* (1995) in the construction of a human YAC contig map.

Restriction enzyme fingerprinting

A physical map of the *E. coli* chromosome

The principle of restriction enzyme fingerprinting was originally developed for the nematode *Caenorhabditis elegans* (Coulson *et al.* 1986) and yeast (Olson *et al.* 1986). However, the first construction of a complete physical map was that of *E. coli*. It is a particularly elegant example and has been selected here for illustrative purposes. The starting point for the *E. coli* map (Kohara *et al*. 1987) was the construction of a genomic library in a lytic phage λ vector. Although cosmid clones contain much longer stretches of cloned DNA they frequently accumulate deletions of *E. coli* DNA because of metabolic imbalances that affect the growth of host cells. These are caused by increased dosage of particular genes on the cosmid. A lytic λ vector has a smaller cloning capacity but has the advantage that the cloned fragments will be stably maintained because it kills host bacteria within a short time.

Once the genomic library had been constructed each clone was individually restriction enzyme fingerprinted. For this purpose, DNA from each clone was partially digested in separate reactions with eight different restriction enzymes. The fragments produced were separated by agarose gel electrophoresis. Each gel was subjected to Southern blotting using as a probe a radioactively-labelled 2.7 kb fragment from the λ vector. The pattern of bands detected was then read in an analogous fashion to a DNA sequencing gel (Fig. 4.2). Starting with 1056 clones, this step was completed in 4 weeks and usable data were generated for 1025 clones.

The restriction fingerprint of each clone then was compared with that of all other clones. Overlapping clones were selected on the basis of a match of at least five consecutive cleavage sites (Fig. 4.2). In this way the 1025 clones were sorted into 70 groups including seven stand-alone clones. The size of the groups ranged from 20 to 180 kb and their sum amounted to 4.4 Mb, i.e. 94% of the genome. Since the average insert size of the library was 15.5 kb, 1025 clones should theoretically represent 96% of the genome (see p. 45). This suggested that the gap between each group could not be larger than several kilobases in length. Thus it should be possible to close each gap with a single clone.

One way of closing the gaps in the map would be to isolate a fresh set of clones and fingerprint them. However, this would be labour intensive. Instead the clones at the ends of each group were used as probes to identify all the other clones with which they would hybridize. Thus clone 6E8, which had been located at the end of group 33, hybridized to clones 22E3 and 14E10. Clone 22E3 also hybridized to clone 9F3 which was located at the end of group 6. Thus clone 22E3 could bridge groups 33 and 6 and this was confirmed by analysis of its restriction fingerprint (Fig. 4.3).

The availability of a complete physical map like that produced by Kohara *et al*. (1987) is a very powerful tool. For example, computer analysis of the nucleotide sequence of a cloned gene allows a simple restriction map to be derived and hence located on the physical map. For *E. coli* sequence data two groups have developed appropriate programs. One is based on restriction fragment alignment (Rudd *et al.* 1990, 1991) and the other on the comparison of the length of each restriction fragment (Médigue *et al.* 1990, 1993).

Other examples of restriction fragment mapping

The fingerprinting technique described above for mapping the *E. coli* genome has been used, albeit with some variation, with the genomes of other microbes (see Cole & Saint Girons 1994 and

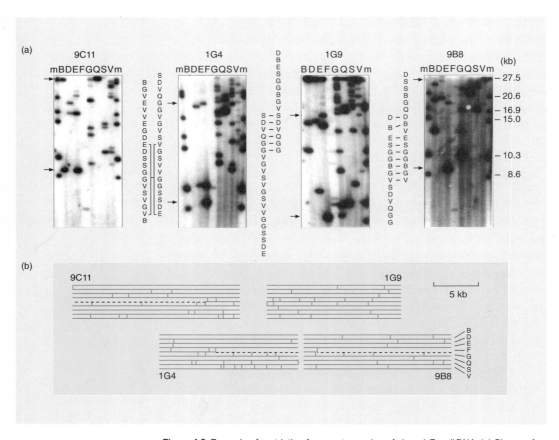

Figure 4.2 Example of restriction fragment mapping of cloned *E. coli* DNA. (a) Clones of *E. coli* DNA were partially digested with restriction enzymes *Bam*HI(B), *Hind*III(D), *Eco*RI(E), *Eco*RV(F), *Bgl*II(G), *Kpn*I(Q), *Pst*I(S) and *Pvu*II(V) and the fragments were separated by electrophoresis. The autoradiograms of four clones are shown after Southern blotting and hybridization with a λ vector DNA probe. In each electropherogram, m represents the lane for the size markers. The boundaries of the cloned DNA and the vector are indicated by arrows. The cleavage sites for each enzyme were 'read' and aligned. The restriction maps deduced from the autoradiograms are depicted in (b). Rightward direction in the maps corresponds to upward (1G4, 1G9 and 9B8) or downward (9C11) in the autoradiograms. Broken lines in the map indicate that the restriction enzyme cleavage sites above them were unable to be read out from the autoradiograms. The right-most *Bam*HI cleavage site found in clone 9C11 was not found in an overlapping clone, 1G4. It is likely that this site had been created in clone 9C11 at the time of library construction by ligating a particular *Sau*3A end such as GATCC into the *Bam*HI site of the vector. (Redrawn with permission from Kohara *et al.* 1987, © Cell Press.)

Fonstein & Haselkorn 1995 for review) as well as eukaryotic organisms, e.g. *Saccharomyces cerevisiae* (Olson *et al.* 1986), *Caenorhabditis elegans* (Coulson *et al.* 1986), *Drosophila melanogaster* (Sidén-Kiamos *et al.* 1990), *Arabidopsis thaliana* (Hauge *et al.* 1991), human chromosome 19 (Trask *et al.* 1992) and the entire human genome (Bellanné-Chantelot *et al.* 1992). The original fingerprinting method devised by Coulson *et al.* (1986) is different from that of Kohara *et al.* (1987) and is shown in Fig. 4.4. Cloned DNA is digested with a restriction endonuclease with a hexanucleotide recognition sequence and which leaves staggered ends, e.g. *Hin*dIII. The ends of the fragments are labelled by end-filling with reverse transcriptase in the presence of a radioactive nucleotide triphosphate. The *Hin*dIII is destroyed by heating and the fragments cleaved again with a restriction enzyme with a tetranucleotide recognition sequence, e.g. *Sau*3A. The fragments then are separated on a high resolution gel and detected by autoradiography. In this case the fingerprint is prepared by determining the size of each band from each clone. The fingerprint obtained by this method is not an order of restriction sites. Rather, it is a series of clusters of bands based on the probability of overlap of clones.

The mapping of the *Drosophila* genome was facilitated by the existence of giant polytene chromosomes in the larval salivary glands. These polytene chromosomes exhibit a pattern of bands and interbands that is remarkably constant. Sidén-Kiamos *et al.* (1990) microdissected discrete and identifiable pieces of these polytene chromosomes and amplified them by PCR to generate probes. These probes then were used to subdivide a genomic library of cosmid clones into families of clones from the same region of the chromosome. The various members of the family then were assembled into contigs on the basis of restriction fragment fingerprints, as described above for *C. elegans*. Once the contig maps had been prepared, Sidén-Kiamos *et al.* (1990)

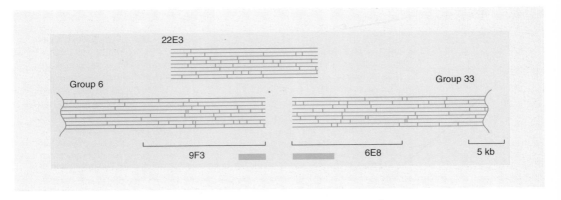

Figure 4.3 Example of the identification of a clone linking two contigs. The positions of probe DNA are indicated by solid bars. See text for details. (Redrawn with permission from Kohara *et al.* 1987, © Cell Press.)

located them physically on the polytene chromosome by *in situ* hybridization.

In preliminary experiments in which DNA was microdissected from *D. melanogaster* chromosomes and used to probe the *D. melanogaster* master library, hybridization with dispersed repetitive DNA proved to be a considerable problem. This is not surprising since about 12% of the *D. melanogaster* genome is estimated to be middle repetitive DNA (see Fig. 2.6). To circumvent this problem probes were prepared from the polytene chromosomes of *Drosophila simulans* which has a much lower content of repetitive DNA and few of its repetitive sequences are found in *D. melanogaster*.

In the whole human genome approach of Bellanné-Chantelot *et*

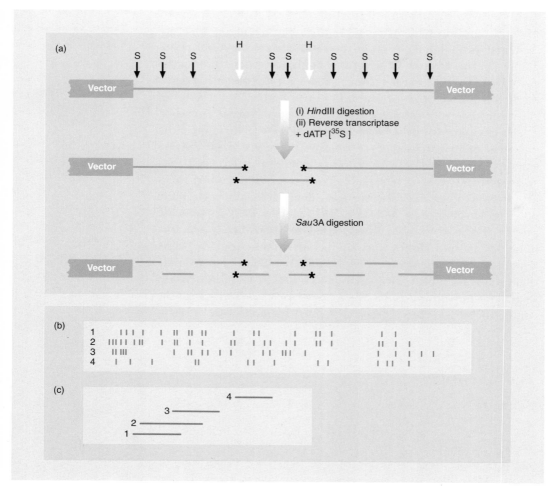

Figure 4.4 The principle of restriction-fragment fingerprinting. (a) The generation of labelled restriction fragments. See text for details. (b) Pattern generated from four different clones. Note the considerable band sharing between clones 1, 2 and 3 indicating that they are contiguous whereas clone 4 is not contiguous and has few bands in common with the other three. (c) The contig map produced from data shown in (b). (Adapted and redrawn with permission from Coulson *et al.* 1986.)

al. (1992) the contigs also were assigned to specific regions of chromosomes by *in situ* hybridization. However, the method used for assembling contigs was different. In the mapping efforts for *E. coli* and *D. melanogaster* described above, the overlap between two clones was detected by preparing a restriction-fragment fingerprint of each clone and identifying restriction-fragment lengths that are common to the two fingerprints. With this method, two clones have to overlap by at least 50% in order to declare with a high degree of certainty that the two clones do indeed overlap. Clearly, increasing the information content in each clone fingerprint would make smaller overlaps detectable. This is particularly important, as Lander and Waterman (1988) showed that the size of the smallest detectable clone overlap is an important parameter in determining the rate at which contigs increase in length. This, in turn, affects the rate at which mapping is completed. For example, the calculated rate of progress increases significantly if the detectable clone overlap is reduced from 50% to 25% of the clone lengths. To enable small overlaps to be detected, Bellanné-Chantelot *et al.* (1992) digested each clone with *Pvu*II, separated the fragments by gel electrophoresis and then prepared a fingerprint by hybridization with a probe prepared from a LINE-1 repeated sequence. Clones that showed little or no hybridization with the LINE-1 probe were fingerprinted with a probe based on an Alu consensus sequence.

There are a number of disadvantages with the fingerprinting approach. First, it is labour intensive and involves extensive handling of clones. Although automation helps, mapping of large genomes still is not an easy task. Second, the method quickly generates a large number of small contigs but it becomes increasingly difficult to extend and join these contigs. Finally, the method works well with phage clones but results with YACs are poor and susceptible to chimaera problems. YACs can be mapped more easily by means of sequence tagged sites.

Marker sequences

Sequence-tagged sites

The concept of sequence-tagged sites (STSs) was developed by Olson *et al.* (1989) in an attempt to systematize landmarking of the human genome. Basically, an STS is a short region of DNA about 200–300 bases long whose exact sequence is found nowhere else in the genome. Two or more clones containing the same STS must overlap and the overlap must include the STS.

Any clone that can be sequenced may be used as an STS provided it contains a unique sequence. A better method to develop STS markers is to create a chromosome-specific library in phage M13. Random M13 clones are selected and 200–400 bases sequenced. The sequence data generated are compared with all known repeated sequences to help identify regions likely to be unique. Two PCR primer sequences are selected from the unique regions which are separated by 100–300 bp and whose melting temperatures are similar (Fig. 4.5). Once identified, the primers are synthesized and used to PCR amplify genomic DNA from the target organism and the amplification products analysed by agarose gel electrophoresis. A functional STS marker will amplify a single target region of the genome and produce a single band on an electrophoretic gel at a position corresponding to the size of the target region (Fig. 4.6). Alternatively, an STS marker can be used as a hybridization probe.

Operationally, an STS is specified by the sequence of the two primers that make its production possible. Thus it is defined once and for all regardless of whether the region under study is cloned in a phage, a cosmid, a BAC, PAC or YAC. Moreover, the STS will remain valid if the corresponding area of the genome is re-cloned sometime in the future in a new and, as yet, unknown vector. The STS is fully portable once the two sequences of the primer are known and these can be obtained from databanks (see Chapter 5).

The principle of the use of STSs to generate physical maps has been confirmed by a number of workers. Thirty YAC clones from the cystic fibrosis region of human chromosome 7 were assembled into a single contig (Green & Olson 1990) that spans more than 1.5 Mb. At the same time, individual YACs as large as 790 kb and containing the entire cystic fibrosis gene were constructed *in vivo* by meiotic recombination in yeast of overlapping YACs. Foote *et al*. (1992) were able to assemble 196 recombinant clones into a single overlapping array which included over 98% of the euchromatic portion of the human Y chromosome. Similarly, Chumakov *et al*. (1992b) were able to generate an STS map for human

Figure 4.5 Example of an STS. The STS developed from the sequence shown above is 171 bases long. It starts at base 162 and runs through base 332. Primer A is 21 bases long and lies on the sequenced strand. Primer B is also 21 bases long and is complementary to the shaded sequence towards the 3′ end of the sequenced strand. Note that the melting temperatures of the two primers are almost equal. (Reproduced with permission from Dogget 1992, courtesy of University Science Books.)

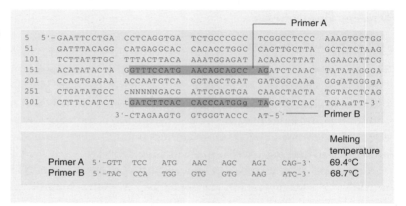

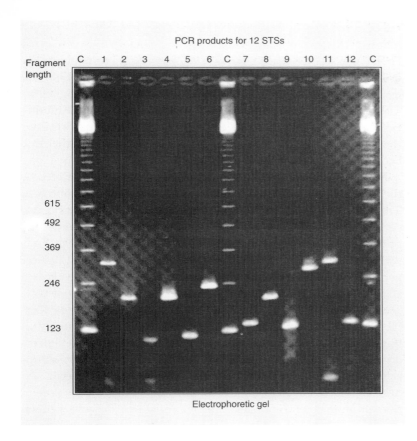

PCR products for 12 STSs

Fragment length

Electrophoretic gel

Figure 4.6 Confirmation that an STS is a unique sequence on the genome. Note that the 12 STSs from chromosome 16 shown above appear as single bands after amplification and hybridization to a chromosome 16 genomic library. (Reproduced with permission from Dogget 1992, courtesy of University Science Books.)

chromosome 21q which was consistent with physical and genetic mapping data.

Radiation hybrid mapping

This method, which has been used only for mapping the human genome, makes use of somatic cell hybrids (see p. 53). A high dose of X-rays is used to break the human chromosome of interest into fragments and these fragments are recovered in rodent cells. The rodent–human hybrid clones are isolated and examined for the presence or absence of specific human DNA markers. The farther apart two markers are on the chromosome, the more likely a given dose of X-rays will break the chromosome between them, placing the markers on two separate chromosomal fragments. By estimating the frequency of breakage, and thus the distance, between markers it is possible to determine their order in a manner analogous to conventional meiotic mapping.

Radiation hybrid mapping was first developed by Goss and Harris (1975). In their experiments, human peripheral blood lymphocytes were irradiated and then fused to hypoxanthine phosphoribosyl transferase (HPRT)-deficient hamster cells. Growth in HAT medium resulted in the isolation of a set of clones, each carrying a different X chromosome fragment that included the selected HPRT marker. Goss and Harris (1977) were able to establish the order of three markers on the long arm of the X chromosome and to demonstrate retention of non-selected chromosome fragments. They also derived mathematical approaches for constructing genetic maps on the basis of co-retention frequencies. However, the power of the technology could not be exploited at the time because insufficient genetic markers were available.

Renewed interest in radiation hybrid mapping was prompted by the work of Cox *et al.* (1990) who modified the original approach. They used as a donor cell a rodent–human somatic cell hybrid that contained a single copy of human chromosome 21 and very little other human DNA. This cell line was exposed to 80.0 Gy of X-ray which resulted in an average of five human chromosome-21 pieces per cell. Because broken chromosomal ends are rapidly healed after X-irradiation, the human chromosomal fragments are usually present as translocations or insertions into hamster chromosomes. However, some cells contain a fragment consisting entirely of human chromosomal material with a human centromere. Since a dose of 80.0 Gy of X-rays results in cell death, the irradiated donor cells were fused with HPRT-deficient hamster recipient cells and hybrids selected as before on HAT medium. Non-selective retention of human chromosomal fragments seems to be a general phenomenon under these fusion conditions. In total, 103 independent somatic cell hybrid clones were isolated and assayed by Southern blotting for the retention of 14 DNA markers. Analysis of the results enabled the 14 markers to be mapped to a 20 Mb region of chromosome 21 and this map order was confirmed by PFGE analysis.

James *et al.* (1994) extended the work of Cox *et al.* (1990) by generating a high resolution radiation hybrid map of human chromosome 11 using 506 STSs scored on a panel of 86 radiation hybrids. A subset of 260 STSs was used to form a map with a resolution of 1 Mb between adjacent positions and which was ordered with odds of 1000 : 1.

As used by Cox *et al.* (1990) and James *et al.* (1994), irradiation and fusion gene transfer is a cumbersome method for creating maps of entire genomes. Barrett (1992) has estimated that between 100 and 200 hybrids are needed to map each chromosome, although James *et al.* (1994) used only 86, and a map of the whole genome would require over 4000 hybrids. This would be both expensive and laborious and would demand the availability of a complete set of human–rodent hybrids containing a single human chromosome. Walter *et al.* (1994) have found a solution to this problem by reverting to the original protocol of Goss and Harris (1975). Instead of using a human–rodent hybrid as a donor, they

CHAPTER 4
*Assembling a physical
map of the genome*

used a diploid human fibroblast. Using 44 radiation hybrids they constructed a map of human chromosome 14 containing 400 ordered markers and concluded that a high resolution map of the whole human genome is feasible with only a single panel of 100–200 radiation hybrids. This subsequently was confirmed by Gyapay *et al.* (1996) who constructed a radiation hybrid map of the human genome using a panel of 168 hybrids.

Work is in progress on a number of mammalian radiation hybrid panels, including rat, mouse, pig and baboon. The technique also is being extended to non-mammalian animals such as zebra fish and chicken (McCarthy 1996).

Expressed sequence tags (ESTs)

In organisms with large amounts of repetitive DNA the generation of an appropriate sequence, and confirmation that it is an STS, can be time consuming. Adams *et al.* (1991) have suggested an alternative approach. The principle of the method is based on the observation that spliced mRNA contains sequences that are largely free of repetitive DNA. Thus partial cDNA sequences, termed ESTs (expressed sequence tags), can serve the same purpose as the random genomic STSs but have the added advantage of pointing directly to an expressed gene. In a test of this concept, partial DNA sequencing was conducted on 600 randomly selected human cDNA clones to generate ESTs. Of the sequences generated, 337 represented new genes, including 48 with similarity to genes from other organisms, and 36 matched previously sequenced human nuclear genes. Forty-six ESTs were mapped to chromosomes.

In practice, there are a number of operational considerations associated with the use of ESTs. First, they need to be very short to ensure that the two ends of the sequence are contiguous in the genome, i.e. are not separated by an intron. Second, large genes may be represented by multiple ESTs which may correspond to different portions of a transcript or various alternatively spliced transcripts. For example, one of the major databases holds over 1300 different EST sequences for a single gene product, serum albumin. While this may or may not be a problem in constructing a physical map, it is problematical in the construction of a genetic map.

If it is desirable to select a single representative sequence from each unique gene, then this is accomplished by focusing on 3′ untranslated regions (3′ UTRs) of mRNAs. This can be achieved using oligo(dT) primers if the mRNA has a poly(A) tail. Two advantages of using the 3′ UTRs are that they rarely contain introns and they usually display less sequence conservation than do coding regions (Makalowski *et al.* 1996). The former feature leads to PCR product sizes that are small enough to amplify. The latter feature makes it easier to discriminate among gene family members that are very similar in their coding regions.

Polymorphic STSs

So far it has been suggested that an STS yields the same product size from any DNA sample. However, STSs can also be developed for unique regions along the genome that vary in length from one individual to another. This variation in length most often occurs because of the presence of microsatellites. In man these usually take the form of CA (or GT) repeats with the dinucleotide being repeated 5–50 times. These sequences are very attractive because they are highly polymorphic, i.e. they will occur as $(CA)_{17}$ in one person, $(CA)_{15}$ in another, and so on. These repeat units are flanked by unique sequences (Fig. 4.7) which can act as primers for the generation of an STS. By definition, such STSs are polymorphic and can be traced through families along with other DNA markers.

Polymorphic STSs are particularly useful. First, because they occur on average every 10 kb, they serve as very useful landmarks. Second, they act as landmarks on both the physical linkage map and the genetic linkage map for each chromosome and provide points of alignment between the different distance scales on these two types of maps. Weissenbach *et al.* (1992) used a total of 814 of such polymorphic STSs to produce a physical map of the human genome. In 1996 a comprehensive genetic map of the human

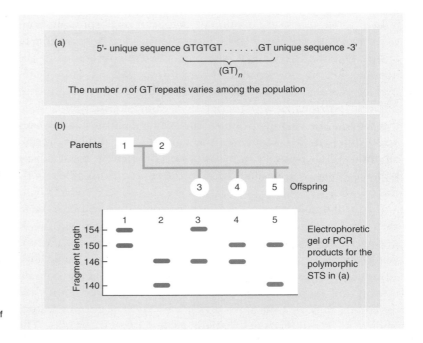

Figure 4.7 The use of a polymorphic STS in inheritance studies. (a) Structure of a polymorphic STS. (b) Schematic representation of a polymorphic STS in a five-member family. The two alleles carried by the father are different from those carried by the mother. The children inherit one allele of the STS from each parent. (Redrawn with permission from Dogget 1992, courtesy of University Science Books.)

genome was completed with the location of 5264 polymorphic STSs at 2335 positions (Dib *et al.* 1996).

Microsatellite markers also have been used to construct a genetic map of the mouse (Dietrich *et al.* 1994, 1996). Because mice can be bred at will it is based on a cross between two species of mice rather than on natural families. The map contains 7377 genetic markers consisting of 6580 highly informative polymorphic STSs integrated with 797 RFLPs, with an average spacing of 0.2 cM (400 kb). In a similar fashion, polymorphic STS maps have been constructed for the pig (Archibald 1994b; Archibald *et al.* 1995), the cow (Barendse *et al.* 1994, Eggen & Fries 1995), the sheep (Crawford *et al.* 1995) and the rat (Serikawa *et al.* 1992; Jacob *et al.* 1995). Polymorphic STSs also have been used extensively in plants (Mazur & Tingey 1995) but are being replaced with amplified fragment length polymorphisms (AFLPs) (see p. 76).

Conventional autoradiographic procedures for analysing amplified microsatellites have limitations in large-scale projects requiring hundreds of thousands of genotypes, particularly since the major source of error is in the reading of autoradiographs. There are, however, alternative methods. Diehl *et al.* (1990) used a laser-based automated DNA fragment analyser to size PCR products from dinucleotide repeats amplified using fluorescently labelled PCR primers. They were able to demonstrate the feasibility of this approach for large-scale human genetic mapping studies. Reed *et al.* (1994) now have produced a set of 254 markers covering all 22 human autosomes and the X chromosome. These markers proved to be highly efficient in the genotyping of microsatellites when incorporated in a standardized fluorescence-based protocol which used a software program to correlate the fluorescence intensity of labelled DNA fragments with marker loci information.

Disadvantages of STSs

Like restriction fragment mapping, STS mapping is still very labour intensive, even with extensive automation. In addition, it is not cheap: to synthesize two primers costs at least $25–50. Thus the cost of generating 100–200 STSs is very significant. However, since gene synthesizers produce much more oligonucleotide than required, different laboratories are sharing their primers. Finally, an STS is defined by the PCR conditions that generate it. Unfortunately, many thermal cyclers used for PCR reactions are not accurately calibrated (Hoelzel 1990), making it difficult to reproduce exactly reaction conditions in different laboratories.

RAPDs and CAPS

Polymorphic DNA can be detected by amplification in the absence of the target DNA sequence information used to generate STSs. Williams *et al.* (1990) have described a simple process, distinct from the PCR process, which is based on the amplification of genomic DNA with *single* primers of arbitrary nucleotide sequence. The nucleotide sequence of each primer was chosen within the constraints that the primer was nine or 10 nucleotides in length, between 50 and 80% G + C in composition and contained no palindromes. Not all the sequences amplified in this way are polymorphic but those that are (randomly amplified polymorphic DNA, RAPD) are easily identified. RAPDs are widely used by plant molecular biologists (Reiter *et al.* 1992; Tingey & Del Tufo 1993) to construct maps because they provide very large numbers of markers and are very easy to detect by agarose gel electrophoresis. However, they have two disadvantages. First, the amplification of a specific sequence is sensitive to PCR conditions, including template concentration, and hence it can be difficult to correlate results obtained by different research groups. For this reason, RAPDs may be converted to STSs after isolation (Kurata *et al.* 1994). A second limitation of the RAPD method is that usually it cannot distinguish heterozygotes from one of the two homozygous genotypes. Nevertheless, Postlethwait *et al.* (1994) have used RAPDs to develop a genetic linkage map of the zebra fish (*Danio rerio*).

A different method for detecting polymorphisms, which is not subject to the problems exhibited by RAPDs, has been described by Konieczny and Ausubel (1993). In this method, STSs are derived from genes which have already been mapped and sequenced. Where possible the primers used are chosen such that the PCR products include introns to maximize the possibility of finding polymorphisms. The primary PCR products are subjected to digestion with a panel of restriction endonucleases until a polymorphism is detected. Such markers are called CAPS (cleaved amplified polymorphic sequences). The way in which CAPS are detected is shown in Fig. 4.8. Note that whereas RFLPs are well suited to mapping newly cloned DNA sequences, they are not convenient to use for mapping genes, such as plant genes, which are first identified by mutation. CAPS are much more useful in this respect.

AFLPs

AFLP is a diagnostic fingerprinting technique that detects genomic restriction fragments and in that respect resembles the RFLP technique (Vos *et al.* 1995). The major difference is that

PCR amplification rather than Southern blotting is used for detection of restriction fragments. The resemblance to the RFLP technique was the basis for choosing the name AFLP. However, the name AFLP should not be used as an acronym because the technique detects *presence* or *absence* of restriction fragments and *not* length differences. The AFLP approach is particularly powerful because it requires no previous sequence characterization of the target genome. For this reason it has been widely adopted by plant geneticists. It also has been used with bacterial and viral genomes (Vos *et al.* 1995). It has not proved useful in mapping animal genomes because it is dependent on the presence of high rates of substitutional variation in the DNA; RFLPs are much commoner in plant genomes compared to animal genomes.

The AFLP technique is based on the amplification of subsets of genomic restriction fragments using PCR (Fig. 4.9). To prepare an AFLP template, genomic DNA is isolated and digested simultaneously with two restriction endonucleases, *Eco*RI and *Mse*I. The former has a 6 bp recognition site and the latter a 4 bp recognition site. When used together these enzymes generate small DNA fragments that will amplify well and are in the optimal size range (< 1 kb) for separation on denaturing polyacrylamide gels. Following heat inactivation of the restriction enzymes the genomic DNA fragments are ligated to *Eco*RI and *Mse*I adapters to generate template DNA for amplification. These common adapter sequences flanking variable genomic DNA sequences serve as primer binding sites on the restriction fragments. Using this strategy it is possible to amplify many DNA fragments without having prior sequence knowledge.

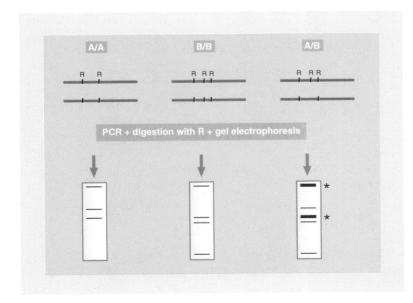

Figure 4.8 Generation and visualization of CAPS markers. Unique-sequence primers are used to amplify a mapped DNA sequence for two different homozygous strains (A/A and B/B) and from the heterozygote A/B. The amplified fragments from strains A/A and B/B contain two and three recognition sites, respectively, for endonuclease R. In the case of the heterozygote A/B, two different PCR products will be obtained, one of which is cleaved twice and the other three times. After fractionation by agarose gel electrophoresis the PCR products from the three strains give readily distinguishable patterns. The asterisks indicate bands that will appear as doublets. (Redrawn with permission from Konieczny & Ausubel 1993.)

The PCR is performed in two consecutive reactions. In the first pre-amplification reaction, genomic fragments are amplified with AFLP primers each having one selective nucleotide (see Fig. 4.9). The PCR products of the pre-amplification reaction are diluted and used as a template for the selective amplification using two new AFLP primers which have two or three selective nucleotides. In addition, the *Eco*RI selective primer is radiolabelled. After the selective amplification the PCR products are separated on a gel and the resulting DNA fingerprint detected by autoradiography (Fig. 4.10).

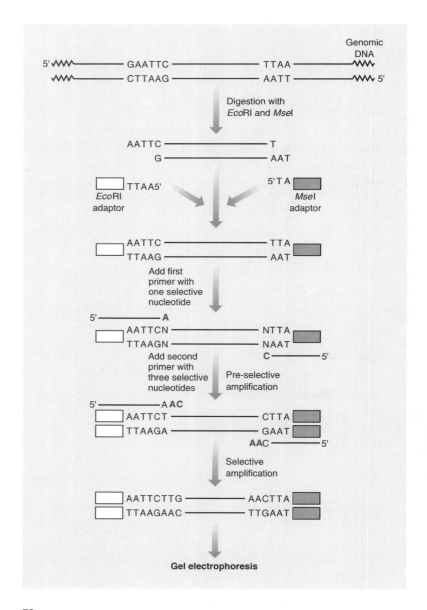

Figure 4.9 Principle of the AFLP method. See text for details.

CHAPTER 4
*Assembling a physical
map of the genome*

The AFLP technique will generate fingerprints of any DNA regardless of the origin or complexity. The number of amplified fragments is controlled by the cleavage frequency of the rare cutter enzyme and the number of selective bases. In addition, the number of amplified bands may be controlled by the nature of the selective bases. Selective extension with rare di- or trinucleotides will result in a reduction of the number of amplified fragments.

The AFLP technique is not simply a fingerprinting technique. Rather, it is an enabling technology that can bridge the gap between genetic and physical maps. Most AFLP fragments correspond to unique positions on the genome and hence can be exploited as landmarks. In higher plants AFLPs may be the most effective way to generate high density maps. The AFLP markers also can be used to detect corresponding genomic clones. Finally, the technique can be used for fingerprinting of cloned DNA segments. By using no or few selective nucleotides, restriction fragment fingerprints will be produced which subsequently can be used to line up individual clones and make contigs.

Genome sequence sampling (GSS)

GSS is a recently described technique (Smith *et al.* 1994) which combines elements of restriction fragment mapping and STS mapping and generates maps with a resolution of 1–5 kb. As

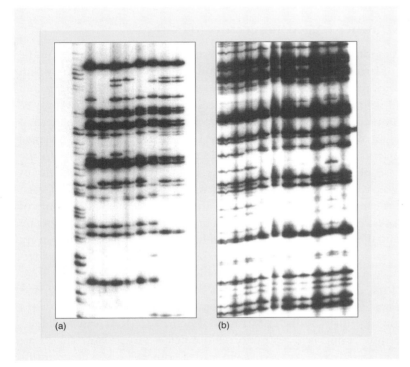

(a)

(b)

Figure 4.10 AFLP analysis of cocoa showing (a) intervarietal variation, and (b) intravarietal variation. (Courtesy of Dr Frances Bligh.)

described above, STSs are used to prepare a physical map of YACs which ideally have been prepared from isolated chromosomes. To produce a high resolution map, a chromosome-specific cosmid library is prepared that represents the genome at 20- to 30-fold redundancy and contains a reasonably random distribution of clone ends. This is produced by cloning using a variety of restriction enzymes and cloning sites. Hybridization of YAC probes to the cosmid clones allows selection of large numbers of those included in the YAC clone. The cosmids are restriction mapped and arranged into contigs as described earlier. At the same time the restriction fragments covering the ends of each cloned piece of DNA are identified by hybridization with purified cosmid vector DNA. Next, automated DNA sequence analysis is carried out using cosmid DNA directly as template and primers recognizing each flanking region of the cosmid vector sequence contiguous to the insert. Thus the sequence of 300–500 bp of each end fragment can be determined with limited accuracy and aligned on the map. If a very high density cosmid map has been prepared, then it enables a high density sequence map to be prepared. For example, Smith *et al.* (1994) studied the protozoan parasite *Giardia lamblia* which has a genome size of 10.5 Mb. A cosmid library of 20-fold redundancy would consist of 5000 unique cosmids and this would generate 10 000 end sequences. Assuming the restriction sites used for cloning were evenly spaced, the DNA sequences determined would be spaced every kilobase, on average.

Sequence-tagged connectors (STCs)

As noted earlier (p. 69) low resolution physical maps can be constructed by using STSs or similar means to order YACs. High resolution maps then can be prepared by randomly cutting and sub-cloning YAC inserts into cosmids which then have to be ordered. Venter *et al.* (1996) have proposed an alternative method involving BACs. A BAC library is constructed with an average insert size of 150 kb and a 15-fold coverage of the genome. In the case of the human genome this would require 300 000 clones which would be arranged in microtitre wells. Both ends of each BAC insert then are sequenced for 500 bases from the point of insert. In the case of the human genome this would generate 600 000 sequences which should be scattered approximately every 5 kb across the genome. These sequences can act as sequence-tagged connectors, or STCs, because they will allow any one BAC clone to be connected to about 30 others since a 150 kb insert 'divided' by 5 kb will be represented in 30 BACs. In this way a physical map can be constructed. Each BAC clone can be finger-printed using one restriction endonuclease to provide the insert size and detect artefactual clones by comparing the fingerprints with those of overlapping clones. It should be noted that this method has not yet been put into practice and Little (1996) has expressed some doubts about its efficiency.

Hybridization assays

Hybridization mapping

This method starts with a genomic library as before. Five kinds of probes, representing known repetitive sequences (centromeric, telomeric, 17S and 5S ribosomal and the long terminal repeat (LTR) of retrotransposons), are hybridized to the library to identify those clones that contain only unique DNA. A number of clones carrying unique DNA are selected at random and used as hybridization probes to detect overlapping clones (see Fig. 4.11). From these clones which do not give a positive hybridization signal another set is selected at random for use as probes in the

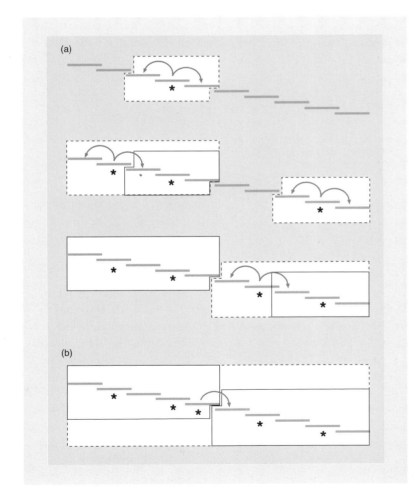

Figure 4.11 The principle of hybridization mapping. (a) Clones for use as probes are randomly picked (*) from a given set of cosmids whose map order is not known. Hybridization identifies overlapping clones (arrows). From clones that do not give a positive signal in any earlier hybridization assay (unboxed areas), probes for the next round of experiments are chosen until all the clones show positive hybridization at least once. (b) Gaps in the map caused by the lack of probes for certain overlap regions are closed by using terminal contig clones. (Redrawn with permission from Hoheisel 1994.)

next round of experiments. This process is continued until all clones show positive hybridization at least once. In practice, some clones containing repetitive DNA have to be used to join contigs. Nevertheless, a key feature of this method is that clones are randomly chosen as probes based on only one criterion — that the clone has not yet given a positive hybridization signal. By this means, large numbers of redundant hybridizations are avoided.

Two refinements of the above process simplify the construction of contigs. First, probes can be prepared from either of the ends of the cloned DNA by using a vector with inward-facing T3 and T7 promoters located at the cloning site (Fig. 4.12). This simplifies contig generation compared with STSs because in the latter case it is hoped that the STS will lie in a region of overlap. However, if the ends of the inserts were sequenced, rather than generating probes, then these sequences could be used as STSs. The use of insert ends as probes eliminates the problem of false positives (i.e. the presence of a hybridization signal between two clones that do not overlap), which can arise due to cross-hybridization between repetitive elements. This is done by demanding that all pairs of cosmid overlaps be reciprocal if both cosmids in the pair are used as probes (Zhang *et al.* 1994). That is, if cosmid *x* hybridizes to

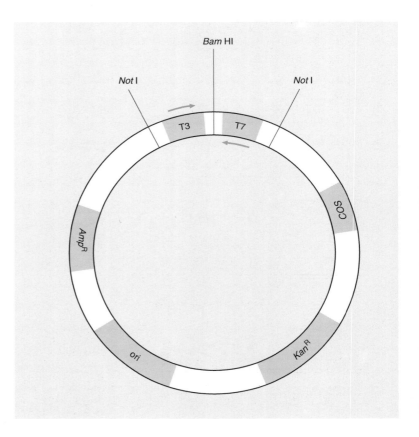

Figure 4.12 A cosmid vector used to generate probes specific for the ends of cloned inserts. The vector contains bacteriophage T3 and T7 promoters flanking a unique *Bam*HI cloning site. Note also the *Not*I sites to facilitate restriction mapping and excision of the insert DNA. ori, origin of replication; *Amp*^R and *Kan*^R, genes conferring resistance to ampicillin and kanamycin respectively; *COS*, cohesive sites essential for *in vitro* packaging in phage λ particles. Arrows show the direction of transcription from the T3 and T7 promoters.

cosmid *y* then *y* must hybridize to *x*. If reciprocity is not achieved, that particular overlap is eliminated from the data set. Second, if a cosmid genomic library and a YAC genomic library are prepared from the same organism, the YAC library can be used to pre-sort the cosmid clones (Fig. 4.13). This mapping procedure can be refined still further, first by using as hybridization probes, short, random-sequence oligonucleotides (Fig. 4.13), then, additional information can be obtained by using as probes sequences representing intron/exon boundaries, or zinc fingers or other structural motifs.

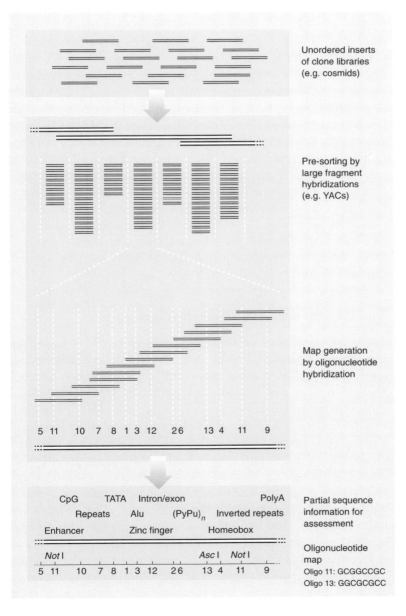

Unordered inserts of clone libraries (e.g. cosmids)

Pre-sorting by large fragment hybridizations (e.g. YACs)

Map generation by oligonucleotide hybridization

5 11 10 7 8 1 3 12 26 13 4 11 9

CpG	TATA	Intron/exon		PolyA	Partial sequence information for assessment
Repeats	Alu	(PyPu)$_n$	Inverted repeats		
Enhancer	Zinc finger	Homeobox			

Not I   *Asc* I   *Not* I

5 11 10 7 8 1 3 12 26 13 4 11 9

Oligonucleotide map
Oligo 11: GCGGCCGC
Oligo 13: GGCGCGCC

Figure 4.13 Principle of oligomer mapping. A high-resolution library (e.g. cosmids) is subdivided by hybridization of low-resolution DNA fragments (e.g. YAC clones). Fingerprinting data for establishing the order of clones are produced by hybridization with short oligonucleotides. Besides providing mapping data, this method yields an oligonucleotide map and partial sequence information concurrently. (Redrawn with permission from Hoheisel 1994.)

The utility of hybridization mapping has been shown by the
construction of a map of the long arm of human chromosome 11
(Evans & Lewis 1989) and a complete map of the fission yeast
(*Schizosaccharomyces pombe*) (Maier *et al*. 1992; Hoheisel *et al*.
1993; Mizukami *et al*. 1993).

Hybridization reference libraries

Once a genomic library has been prepared, then, regardless of the
vector used, it can be kept as a collection of individual clones in
microtitre dishes. Each clone of the library then is spotted as a
regular and reproducible array on a filter, 10 or 20 000 at a time.
Batches of filters are distributed to other laboratories who screen
them with probes generated by them from genes or regions of
interest. Since a few standard-size filters containing an entire
library can be probed multiple times, this is not a major undertak-
ing. When positive hybridization signals are obtained, the location
of the signal and the probe used are notified to the library holder,
who, in turn, provides a sample of the relevant clone(s) (Lehrach
et al. 1990; Zehetner & Lehrach 1994). Thus one library can serve
multiple users while, at the same time, enabling all data to be
centralized to facilitate map generation (Fig. 4.14). Hybridization
libraries are the antithesis of the STS approach: one uses clones as
the reference objects, the other STSs. Similarly, one uses hybridization for screening, the other PCR. Whereas reproducible PCR
conditions are essential for STS mapping, the library approach
allows different hybridization conditions to be used by different
groups.

The importance of *in situ* hybridization

There are a number of different kinds of genome maps, e.g.
cytogenetic, linkage, physical, etc. The classic cytogenetic map
gives visual reality to other maps and to the chromosome itself.
Because it does not rely on the cloning of DNA fragments it
avoids the pitfalls that this procedure can introduce, particularly
with YACs (see p. 57). Genetic linkage mapping allows the localization of inherited markers relative to each other. As with
cytogenetic maps, linkage maps examine chromosomes as they are
in cells. Although the methodology used to construct cytogenetic
and linkage maps can lead to errors, they nevertheless are used as
gold standards against which the physical maps are judged. The
importance of *in situ* hybridization is that it enables this comparison to be made. Providing hybridization of repeated sequences is
suppressed and provided no DNA chimaeras are present, a cloned
fragment or restriction fragment should anneal to a single location
on the cytogenetic map. Furthermore, the physical map order

84

should match that found by *in situ* hybridization. Where genetic markers have been located on the cytogenetic map by *in situ* hybridization they also can be positioned on the physical map.

Originally, *in situ* hybridization of unique sequences utilized radioactively labelled probes and it was a technique which required

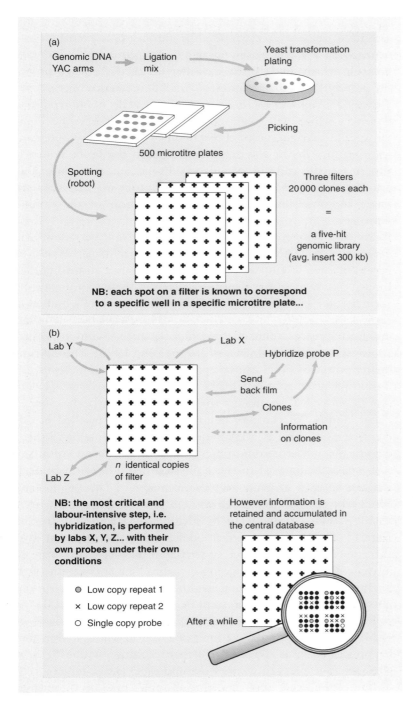

Figure 4.14 The concept of the reference library. (a) A YAC library containing about 50 000 clones stored in 500 microtitre plates. Each of these clones is individually and accurately positioned on the filters in a very fine grid (20 000 clones per filter) so that each corresponds to a defined well, in a particular microtitre plate. (b) The procedure for utilizing the reference library. Sets of identical filters are produced and sent to the laboratories wishing to isolate the YAC(s) corresponding to their probe(s). The laboratories then hybridize the filters and notify the central laboratory of the position of the positive clone(s) which can then be provided. The lower right half of (b) shows how information on each of the library's clones progressively and naturally builds up. (Redrawn with permission from Jordan 1993.)

a great deal of technical dexterity. Today, the methods used are all derivatives of the fluorescence *in situ* hybridization (FISH) method developed by Pinkel *et al.* (1986). In this technique, the DNA probe is labelled by addition of a reporter molecule. The probe is hybridized to a preparation of metaphase chromosomes which has been air-dried on a microscope slide and in which the DNA has been denatured with formamide. Following hybridization, and washing to remove excess probe, the chromosome preparation is incubated in a solution containing a fluorescently labelled affinity molecule which binds to the reporter on the hybridized probe. The preparation is then examined with a fluorescence microscope. If large DNA probes are used, they will contain many repetitive sequences which will bind indiscriminately to the target. This non-specific binding can be eliminated by competitive suppression hybridization. Before the main hybridization the probe is mixed with an aqueous solution of unlabelled total genomic DNA. This saturates the repetitive elements in the probe so that they no longer interfere with the *in situ* hybridization of the unique sequences.

Extensive cytogenetic maps exist for *Drosophila*, because of the presence of polytene chromosomes, and for humans because of the importance of clinical genetics. Thus it is not surprising that detailed *in situ* maps have been prepared for these two organisms (Lichter *et al.* 1990; Hartl *et al.* 1992). In the case of a human, a whole series of different diagnostic probes are available, including ones for telomeres, centromeres, unique sequences and individual chromosomes (van Ommen *et al.* 1995; Yung 1996). The discriminatory power of FISH can be enhanced by the simultaneous use of multiple colours in one hybridization. Dauwerse *et al.* (1992) have developed a system for the simultaneous hybridization of a large number of chromosome-specific libraries, labelled in different ratios of three primary fluorochromes, which allowed them to paint metaphase chromosomes in up to 12 different colours.

As well as being used to check physical maps, *in situ* hybridization can be used to generate physical maps. In *Drosophila*, YAC clones typically hybridize with enough polytene bands that overlapping clones (contigs) can be identified cytologically. In this context, cytological contigs are defined by the rule that two adjacent YACs are considered to overlap if they have two or more polytene chromosome bands in common. The application of *in situ* hybridization to mapping in other organisms is much more difficult. Conventional fluorescent *in situ* hybridization (FISH) as applied to metaphase chromosomes has a resolving power of ~1 Mb. This is the upper size limit of YACs so will not allow fine mapping of cosmids. To achieve higher resolution with FISH it is necessary to use less condensed chromosomes as the target.

Highly elongated metaphase chromosomes have been prepared

CHAPTER 4
*Assembling a physical
map of the genome*

by mechanically stretching them by cytocentrifugation. This results in chromosomes that are 5–20 times their normal length. Laan *et al.* (1995) have shown that these stretched chromosomes are excellent for fast and reliable ordering of clones that are separated by at least 200 kb. They also can be used to establish the centromere–telomere orientation of a clone. The disadvantages of the method are that it cannot be used to generate reliable measurements of the distances between signals nor can it be used to localize unknown sequences on a chromosome. This is because the stretching of individual chromosomes is highly variable.

In interphase nuclei the chromatin is less condensed than in metaphase chromosomes and hence provides a good target for high resolution FISH. Using cosmid clones, Trask *et al.* (1989) showed that FISH to interphase nuclei can be used to determine the genomic distance between the probes over a range of 25–250 kb. More recently they have extended this to distances as large as 2 Mb (Yokota *et al.* 1995).

The resolution of FISH can be further improved by loosening the organization of the interphase chromatin using high salt, alkali or detergent treatment of the cell preparations (Parra & Windle 1993). These techniques, which are commonly referred to as fibre-FISH, provide a resolution which permits the detection of a probe to a single DNA fibre. Fibre-FISH allows the orientation of clones and *in situ* mapping at distances from 1–500 kb, the assessment of cosmid and YAC overlaps, the sizing of uncloned gaps between adjacent contigs, the assessment of linear continuity of YACs in deletion prone genomic regions, and the rapid mapping of cosmids and cDNAs along cloned YACs (Rosenberg *et al.* 1995). Theoretically the resolution of fibre-FISH is the same as the resolving power of the light microscope, i.e. $0.34 \mu m$. This is equivalent to 1 kb and has been achieved by Florijn *et al.* (1995).

A new variation on multicolour FISH is the use of padlock probes (Nilsson *et al.* 1997). Padlock probes are oligonucleotides that can be ligated to form a circle if they bind to a sequence of exact complementarity. The lateral arms of the oligonucleotide twist around the DNA target forming a double helix and their termini are designed to juxtapose so that they may be ligated enzymatically. Nilsson *et al.* (1997) used two different oligonucleotide probes, each corresponding to a different sequence variant of a centromeric alpha-satellite repeat, and differing by only a single base pair. The closure of the two alternative padlocked probes occurred only when there was perfect sequence recognition. By labelling the two probes with different fluorescent dyes it was possible to monitor the two sequence variants simultaneously. Lizard and Ward (1997) have speculated on the possible applications of padlocked probes in genome analysis.

CHAPTER 4
*Assembling a physical
map of the genome*

The advantages and disadvantages of the different methods of FISH are shown in Table 4.1.

Computation and automation

Regardless of which method of physical mapping is used, the ordering of clones into contigs requires that an extensive amount of data has to be processed. Consider the experiments described on p. 64 for the mapping of the *E. coli* genome. A restriction enzyme fingerprint was prepared for 1025 clones and then all had to be compared pairwise to detect overlaps (see Fig. 4.2). This is a mammoth task which is made even greater if the distance between restriction sites has to be included in the analysis, as happened in the generation of the nematode and *Drosophila* maps. Thus it is not surprising that all the mapping methods make extensive use of computers and specially constructed algorithms. Over the years a number of different algorithms have been developed for contig generation using either restriction enzyme fingerprinting (Sulston *et al.* 1988; Branscomb *et al.* 1990) or hybridization methods (Mott *et al.* 1993; Wang *et al.* 1994; Zhang *et al.* 1994). The paper of Zhang *et al.* (1994) presents a particularly clear exposition of the principles of algorithm construction.

Much of the work involved in physical mapping is repetitive and monotonous. Surprisingly, there still is relatively little automation and the current status has been reviewed by Hodgson (1994). Whereas some steps are difficult to automate satisfactorily, e.g. isolation of ultrapure DNA, others such as Southern blotting are not. Similarly, some of the mapping techniques are compatible with automation, e.g. preparation of hybridization libraries,

Table 4.1 Comparison of mapping techniques using FISH. (Reproduced with permission from Heiskanen *et al.* 1996)

Visual mapping of genes and DNA-clones on		Resolution	Application	Advantages (+) and disadvantages (−)
Metaphase chromosomes		> 1 Mb	Chromosome assignment; detection of chimaerism or homologous regions	+ Telomere–centromere orientation + Mapping to specific chromosomal band − Resolution seldom sufficient for clone ordering
Mechanically stretched chromosome		> 200 kb	Ordering of clones	+ Telomere–centromere orientation − Distance determination not possible
Interphase nuclei		~50 kb–1 Mb	Ordering of clones	+ Distance determination possible − Telomere–centromere orientation not possible
DNA fibres		~1 kb–500 kb	Ordering of clones	+ Accurate distance determination − Telomere–centromere orientation not possible

whereas others are not. Whenever possible, short cuts are used to reduce the number of repetitive steps. As just one example, consider hybridization mapping of cosmid clones described on p. 81. Starting from 96 well archive plates, cosmid clones are inoculated on the surface of a filter. Each clone on the grid is assigned a unique identifying Y- and X-axis coordinate. To enable analysis of multiple clones simultaneously, cosmids are pooled according to the rows and columns of the matrix, DNA is prepared and a mixed RNA probe is synthesized. When hybridized to the matrix filter, the probe detects a pattern of spots consisting of all the template clones and the collection of clones overlapping with one end of each of the template clones. A similar procedure is carried out by using cosmids pooled according to columns of the matrix (Evans & Lewis 1989). When the two sets of data are compared, hybridizing clones identified by both of the mixed probes may be overlapping with the template clone common to both sets: the clone located at the intersection of the row and column (Fig. 4.15). It should be noted that with this method, it is essential to eliminate hybridization due to repeat sequences. This is achieved by pre-hybridizing the cosmids with genomic DNA to low C_0t values (see p. 19).

In generating an STS-based map of the human genome Hudson *et al.* (1995) had to process more than 15 million reactions. Automation was the only way this could be achieved and various special-purpose machines were constructed. PCR reactions were carried out in custom-fabricated 1536-well microtitre cards and the cards were incubated in a series of temperature-controlled chambers. After incubation the reaction mixtures were transferred onto a hybridization membrane affixed to the bottom of a second microtitre card. This was achieved by piercing the first card with a bed of 1536 hypodermic needles and sucking the reactions downwards with a vacuum plenum. These filter cards then were hybridized with a chemiluminescent probe and read by a CCD camera. All the activities were computer controlled and the microtitre cards were assigned a barcode to facilitate sample tracking. Overall the throughput was 150 000 reactions per run!

A novel approach to generating maps: optical mapping

From the foregoing it should be clear that the construction of maps for eukaryotic chromosomes is laborious and difficult. Much of the problem derives from the fact that many of the procedures used for mapping and sequencing DNA were designed originally to analyse genes rather than genomes. The electrophoretic methods that are widely used in mapping offer the advantage of good size resolution, even for larger molecules, but require preparation of DNA in bulk amounts from sources such as genomic DNA or

YACs. In contrast, single-molecule techniques such as FISH make use of only a limited number of chromosomes but do not have good size resolution. Ideally, one would like to be able to combine the sizing power of electrophoresis with the intrinsic capability of FISH. Schwartz *et al.* (1993) have developed a technique, optical mapping, which approaches this ideal by imaging single DNA molecules during restriction enzyme digestion.

In practice, a fluid flow is used to stretch out fluorescently stained DNA molecules dissolved in molten agarose and fix them in place during gelation. A restriction enzyme is added to the molten agarose–DNA mixture. Cutting is triggered by the diffusion of magnesium ions into the gelled mixture which has been mounted on a microscope slide. Fluorescence microscopy is used to record at regular intervals the cleavage sites. These are visualized

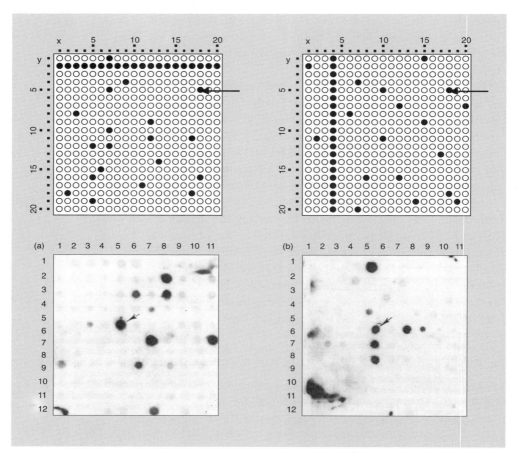

Figure 4.15 Strategy for analysis of physical linkage by using groups of cosmids (see text for details). (a) Schematic representation of the method. The arrows indicate the location of a clone that overlaps with the clone on coordinates y = 2, x = 4. (b) Actual results obtained using this procedure. (Reproduced with permission from Evans & Lewis 1989.)

by the appearance of growing gaps in imaged molecules and bright condensed pools of DNA on the fragment end flanking the cut site. The size of the resulting individual restriction fragments is determined by relative fluorescence intensity and apparent molecular contour length. Wang *et al.* (1995b) have extended the technique to incorporate the use of RecA-assisted restriction endonuclease (RARE) cleavage (see p. 48) and Yokota *et al.* (1997) have developed a new method for straightening the DNA molecules prior to restriction mapping.

Meng *et al.* (1995) improved the size resolution of optical mapping and accurately determined the mass of restriction fragments 800 bp long and Samad *et al.* (1995) reported that the resolution had been improved to 300 bp. Sizing accuracy is a function of the number of molecules analysed and accuracy to within a few base pairs on kilobase-range restriction digest fragments of phage λ has been achieved. This represents higher accuracy than is achievable by agarose gel electrophoresis.

To date, optical mapping has only been used in one laboratory and it remains to be seen if it will be more widely adopted. So far, it has been used to generate ordered restriction maps of YACs, the human Beckwith–Wiedeman and breast cancer *Brca* 2 loci and the mouse olfactory locus (Samad *et al.* 1995). These maps have been constructed from phage, cosmid, YAC and BAC clones. Major advantages of the method are avoidance of the need for gel electrophoresis or extensive computation.

Integration of different mapping methods and measuring progress

Each of the mapping methods described above has its advantages and disadvantages and no one method is ideal. The size and complexity of the genome being analysed can greatly influence the methodologies employed in map construction. Nor is it uncommon for different research groups to use different mapping methods even when working on the same genome. Ultimately the maps generated by the different methods need to be integrated. Also, it is essential to be able to monitor progress to distinguish those genomic regions requiring additional work and resources from those that essentially are complete. Cox *et al.* (1994) have provided a methodology for doing this.

Most genomic mapping projects involve ordering two classes of objects relative to one another. These are *breakpoints* and *markers* (Table 4.2). Breakpoints, so called because they represent subdivisions of the genome, are defined by a specific experimental resource. Markers consist of unique sites in the genome and should be independent of any particular experimental resource. Although both types of objects are essential for map construction, the map itself should be defined in terms of markers, especially those based on DNA sequence. One reason for this is that markers are permanent and easily shared. They can be readily stored as DNA sequence information and distributed in this fashion. By contrast, breakpoints are defined by experimental resources that tend to be transient and cumbersome to distribute. The most important reason to use sequence-based markers is that they can be easily screened against any DNA source. Thus they can be used to integrate maps constructed by diverse methods and investigators. Such integration is crucial to the assembly and assessment of maps. A good example is provided by the increasingly detailed maps of the human genome being produced jointly by a number of large, independent groups (Murray *et al.* 1994; Hudson *et al.* 1995; Schuler *et al.* 1996).

The breakpoints divide the genome into 'bins' corresponding to the regions between breakpoints. In assessing mapping progress a first step is to determine the number of 'bins' that are occupied by markers and the distribution of these markers within each bin. Although some investigators report only the total number of markers used to construct the map, it is the number of occupied bins that provides the measure of progress. The distribution of the number of markers per bin is important because the goal is to have markers evenly, or at least randomly, spread rather than clustered.

The second step in assessing progress is to identify those occupied bins that are ordered relative to one another and to estimate the confidence in the ordering. Note that assignment of markers to bins can proceed throughout a mapping project but the ordering of bins is only possible as a project matures. Thus ordering is a good indication of the degree of completion. Finally, the distance in kilobases between ordered markers in a map needs to be measured.

Synthesis: assembling all the information

As noted in Chapter 1, there are two kinds of maps: genetic maps and physical maps. The former are used to map genes of interest by analysis of the progeny derived from genetic crosses. RFLPs,

Table 4.2 Categorization of map objects. (Reproduced with permission from Cox *et al.* 1994, © American Association for the Advancement of Science)

Mapping method	Experimental resource	Breakpoints	Markers
Meiotic	Pedigrees	Recombination sites	DNA polymorphisms
Radiation hybrid	Hybrid cell lines	Radiation-induced chromosome breaks	STSs
In situ hybridization	Chromosomes	Cytological landmarks	DNA probes
STS content	Library of clones	End points of clones	STSs
Clone-based fingerprinting	Library of clones	End points of clones	Genomic restriction sites

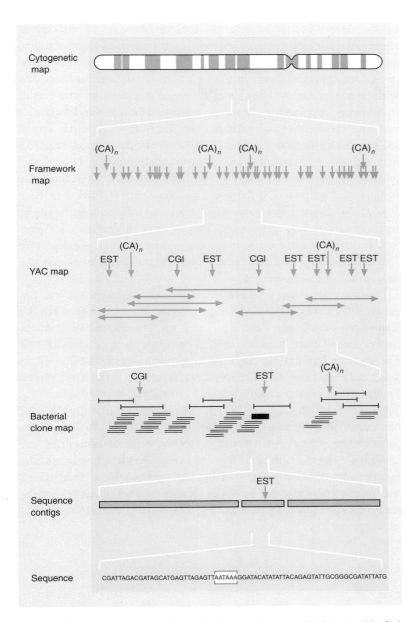

Cytogenetic map

Framework map

$(CA)_n$ $(CA)_n$ $(CA)_n$ $(CA)_n$

YAC map

$(CA)_n$

EST   CGI   EST   CGI   EST EST   EST EST

Bacterial clone map

CGI   EST $(CA)_n$

Sequence contigs

EST

Sequence

CGATTAGACGATAGCATGAGTTAGAGTTAATAAAGGATACATATATTACAGAGTATTGCGGGCGATATTATG

Figure 4.16 The hierarchy of resolution of the integrated chromosome map. At the lowest level of resolution is the cytogenetic map based on the banding patterns of metaphase chromosomes. At the next level lies the framework map, with a series of anchor markers (indicated by vertical arrows) which may be expressed sequence tags (ESTs), CpG island (CGI) sequences, microsatellite repeats (indicated by $[CA]_n$) or anonymous DNA markers, all of which may be placed by PCR on to the radiation hybrid map. The expected density of markers could reach as high as one marker per 100 kb with genetic markers at one per cM. The framework map anchors the YAC map through the markers as shown on the next level of resolution. Additional markers which are not resolvable in the framework map could push the landmark density to one per 50 kb. The YAC map can then be used in combination with fingerprinting to assist the assembly of a bacterial clone map consisting of cosmids (the plain lines) and bacterial or P1 artificial chromosomes (indicated by the longer lines). This has sometimes been termed the 'sequence-ready' map and may have a resolution of < 10 kb. Finally, the bacterial clones may be used to determine the highest resolution physical map, the DNA sequence. The cosmid in bold has been assembled into three sequence contigs via automated shotgun DNA sequencing. The original landmarks are absolutely placed by sequence matching, in this case an EST showing an exact match with the DNA sequence. Lower levels of similarity, as well as other gene detection methods, will eventually place all the genes within the DNA sequence. (Reproduced with permission from Bentley & Dunham 1995, © *Current Opinion in Genetics and Development*.)

CAPS, AFLPs and polymorphic STSs are examples of the molecular markers which can be used in the generation of genetic maps. To be maximally useful such markers must have high heterozygosity. In other words, there should be a high likelihood that a marker is different in any two copies of a chromosome. Also, the higher the density of markers the more useful the map will be. In the current version of the genetic map of the human genome (Dib *et al*. 1996) 5264 polymorphic STSs have been located at 2335 positions. The markers have an average heterozygosity of 0.7 which is a very useful level and about the best that can realistically be achieved. On average there is a marker every 0.7 cM, i.e. there is a 0.7% chance that any two markers will be separated by recombination during meiosis. This density of markers is higher than originally set for the mapping of the human genome (see Table 1.2) and is of a level that will facilitate the identification of new genes (see Chapter 7).

Physical maps are a prerequisite for genomic sequencing. Because only short lengths of DNA (typically 500 bases) can be sequenced in a single step, a physical map needs to have a much higher density of markers than a genetic map. Thus, for the purposes of sequencing the human genome, work is ongoing to generate a physical map in which 30 000 STS markers are placed on the genetic map of Dib *et al*. (1996) at an average density of 100 kb (Jordan & Collins 1996). Already preliminary maps have been published (Hudson *et al*. 1995; Schuler *et al*. 1996). In the most recent version over 20 000 STSs had been mapped. The way in which a detailed physical map can be constructed by synthesizing the information generated by different mapping methods is shown schematically in Fig. 4.16. Details of a real example can be found in the review of Bentley and Dunham (1995).

5 Sequencing methods and strategies

Basic DNA sequencing

The first significant DNA sequence to be obtained was that of the cohesive ends of phage λ DNA (Wu & Taylor 1971) which are only 12 bases long. The methodology used was derived from RNA sequencing and was not applicable to large-scale DNA sequencing. An improved method, plus and minus sequencing, was developed and used to sequence the 5386 bp phage ΦX 174 genome (Sanger *et al.* 1977a). This method was superseded in 1977 by two different methods, that of Maxam and Gilbert (1977) and the chain termination or dideoxy method (Sanger *et al.* 1977b). For a while the Maxam and Gilbert method, which makes use of chemical reagents to bring about base-specific cleavage of DNA, was the favoured procedure. However, refinements to the chain-termination method meant that by the early 1980s it became the preferred procedure. To date, most large sequences have been determined using this technology, with the notable exception of bacteriophage T7 (Dunn & Studier 1983). For this reason, only the chain-termination method will be described here.

The chain terminator or dideoxy procedure for DNA sequencing capitalizes on two properties of DNA polymerases: (i) their ability to synthesize faithfully a complementary copy of a single-stranded DNA template; and (ii) their ability to use $2',3'$-dideoxynucleotides as substrates (Fig. 5.1). Once the analogue is incorporated at the growing point of the DNA chain, the $3'$ end lacks a hydroxyl group and no longer is a substrate for chain elongation. Thus, the growing DNA chain is terminated, i.e. dideoxynucleotides act as chain terminators. In practice, the Klenow fragment of DNA polymerase is used because this lacks the $5' \rightarrow 3'$ exonuclease activity associated with the intact enzyme. Initiation of DNA synthesis requires a primer and usually this is a chemically synthesized oligonucleotide which is annealed close to the sequence being analysed.

DNA synthesis is carried out in the presence of the four deoxynucleoside triphosphates, one or more of which is labelled with ^{32}P, and in four separate incubation mixes containing a low concentration of one each of the four dideoxynucleoside triphosphate analogues. Therefore, in each reaction there is a population of partially synthesized radioactive DNA molecules, each having a common 5′-end, but each varying in length to a base-specific 3′ end (Fig. 5.2). After a suitable incubation period, the DNA in each mixture is denatured and electrophoresed in a sequencing gel.

A sequencing gel is a high-resolution gel designed to fractionate single-stranded (denatured) DNA fragments on the basis of their size and which is capable of resolving fragments differing in length by a single base pair. They routinely contain 6–20% polyacrylamide and 7 M urea. The function of the urea is to minimize DNA secondary structure which affects electrophoretic mobility. The gel is run at sufficient power to heat up to about 70°C. This also minimizes DNA secondary structure. The labelled DNA bands obtained after such electrophoresis are revealed by autoradiography on large sheets of X-ray film and from these the sequence can be read (Fig. 5.3).

To facilitate the isolation of single strands the DNA to be sequenced may be cloned into one of the clustered cloning sites in

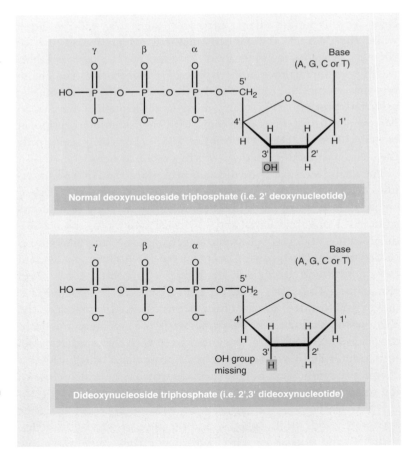

Figure 5.1 Dideoxynuclcoside triphosphates act as chain terminators because they lack a 3′-OH group. Numbering of the carbon atoms of the pentose is shown (primes distinguish these from atoms in the bases). The α, β and γ phosphorus atoms are indicated.

the *lac* region of the M13 mp series of vectors (Fig. 5.4). A feature of these vectors is that cloning into the same region can be mediated by any one of a large selection of restriction enzymes but still permits the use of a single sequencing primer.

Modifications of chain terminator sequencing

The sharpness of the autoradiographic images can be improved by

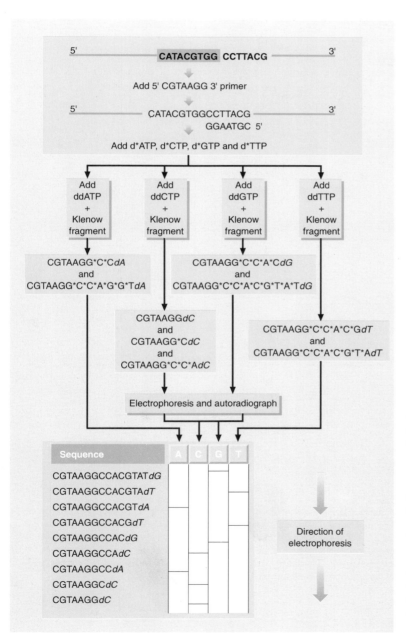

Figure 5.2 DNA sequencing with dideoxynucleoside triphosphates as chain terminators. In this figure asterisks indicate the presence of ^{32}P and the prefix 'd' indicates the presence of a dideoxynucleoside. At the top of the figure the DNA to be sequenced is enclosed within the box. Note that unless the primer is also labelled with a radioisotope the smallest band with the sequence CGTAAGG*dC* will not be detected by autoradiography as no labelled bases were incorporated.

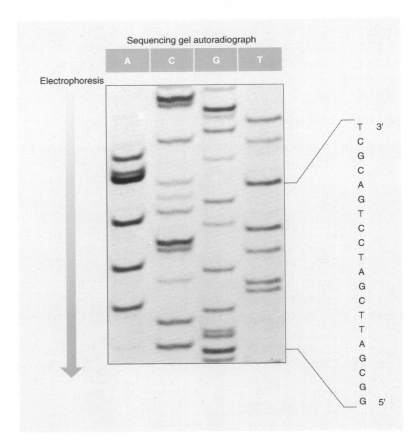

Figure 5.3 Enlarged autoradiograph of a sequencing gel obtained with the chain terminator DNA sequencing method.

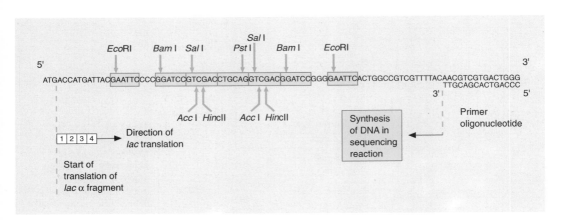

Figure 5.4 Sequence of M13 mp7 DNA in the vicinity of the multipurpose cloning region. The upper sequence is that of M13 mp7 from the ATG start codon of the β-galactosidase α-fragment, through the multipurpose cloning region, and back into the β-galactosidase gene. The short sequence at the right-hand side is that of the primer used to initiate DNA synthesis across the cloned insert. The numbered boxes correspond to the amino acids of the β-galactosidase fragment.

replacing the ^{32}P-radiolabel with the much lower energy ^{33}P or ^{35}S. In the case of ^{35}S, this is achieved by including an α-^{35}S-deoxynucleoside triphosphate (Fig. 5.5) in the sequencing reaction. This modified nucleotide is accepted by DNA polymerase and incorporated into the growing DNA chain. Non-isotopic detection methods also have been developed with chemiluminescent, chromogenic or fluorogenic reporter systems. Although the sensitivity of these methods is not as great as with radiolabels, it is adequate for many purposes.

Other technical improvements to Sanger's original method have been made by replacing the Klenow fragment of *E. coli* DNA polymerase I. Natural or modified forms of the phage T7 DNA polymerase ('Sequenase') have found favour, as has the DNA polymerase of the thermophilic bacterium *Thermus aquaticus* (Taq DNA polymerase). The T7 DNA polymerase is more processive than Klenow polymerase, i.e. it is capable of polymerizing a longer run of nucleotides before releasing them from the template. Also, its incorporation of dideoxynucleotides is less affected by local nucleotide sequences and so the sequencing ladders comprise a series of bands with more even intensities. The Taq DNA polymerase can be used in a chain termination reaction carried out at high temperatures (65–70°C) and this minimizes chain termination artefacts caused by secondary structure in the DNA. Tabor and Richardson (1995) have shown that replacing a single phenylalanine residue of Taq DNA polymerase with a tyrosine residue results in a thermostable sequencing enzyme that no longer discriminates between dideoxy- and deoxynucleotides.

The combination of chain terminator sequencing and M13 vectors to produce single-stranded DNA is very powerful. Very good quality sequencing is obtainable with this technique, especially when the improvements given by ^{35}S-labelled precursors and T7 DNA polymerase are exploited. Further modifications allow sequencing of 'double-stranded' DNA, i.e. double-stranded input DNA is denatured by alkali, neutralized, and one strand then is

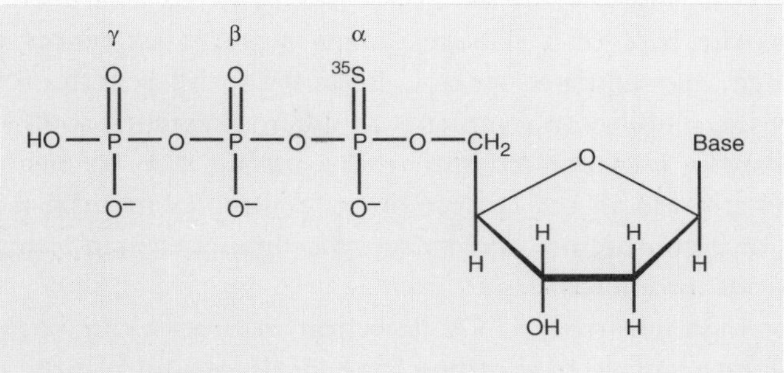

Figure 5.5 Structure of an α-^{35}S-deoxynucleoside triphosphate.

annealed with a specific primer for the actual chain terminator sequencing reactions. This approach has gained in popularity as the convenience of having a universal primer has grown less important with the widespread availability of oligonucleotide synthesizers. With this development, Sanger sequencing has been liberated from its attachment to the M13 cloning system; for example, PCR-amplified DNA segments can be sequenced directly. One variant of the double-stranded approach, often employed in automated sequencing, is 'cycle sequencing'. This involves a *linear* amplification of the sequencing reaction using 25 cycles of denaturation, annealing of a specific primer to one strand only, and extension in the presence of Taq DNA polymerase plus labelled dideoxynucleotides. Alternatively, labelled primers can be used with unlabelled dideoxynucleotides.

Automated DNA sequencing

In manual sequencing, the DNA fragments are radiolabelled in four chain termination reactions, separated on the sequencing gel in four lanes, and detected by autoradiography. This approach is not well suited to automation. To automate the process it is desirable to acquire sequence data in real-time by detecting the DNA bands within the gel during the electrophoretic separation. However, this is not trivial as there are only about 10^{-15}–10^{-16} moles of DNA per band. The solution to the detection problem is to use fluorescence methods. In practice, the fluorescent tags are attached to the chain-terminating nucleotides. Each of the four dideoxynucleotides carries a spectrally different fluorophore. The tag is incorporated into the DNA molecule by the DNA polymerase and accomplishes two operations in one step: it terminates synthesis and it attaches the fluorophore to the end of the molecule. Alternatively, fluorescent primers can be used with non-labelled dideoxynucleotides. By using four different fluorescent dyes it is possible to electrophorese all four chain-terminating reactions together in one lane of a sequencing gel. The DNA bands are detected by their fluorescence as they electrophorese past a detector (Fig. 5.6). If the detector is made to scan horizontally across the base of a slab gel, many separate sequences can be scanned, one sequence per lane. Because the different fluorophores affect the mobility of fragments to different extents, sophisticated software is incorporated into the scanning step to ensure that bands are read in the correct order. A simpler method is to use only one fluorophore and to run the different chain-terminating reactions in different lanes.

For high-sensitivity DNA detection in four-colour sequencing and high-accuracy base calling, one ideally would like the following criteria to be met: each of the four dyes to exhibit strong absorption at a common laser wavelength; to have an emission maximum at a distinctly different wavelength; and to introduce the same relative mobility shift of the DNA sequencing fragments. Recently, dyes with these properties have been identified and successfully applied to automated sequencing (Glazer & Mathies 1997).

Automated DNA sequencers offer a number of advantages that are not particularly obvious. First, manual sequencing can generate excellent data but even in the best sequencing laboratories poor autoradiographs frequently are produced that make sequence reading difficult or impossible. Usually the problem is related to the need to run different termination reactions in different tracks of the gel. Skilled DNA sequencers ignore bad sequencing tracks but many laboratories do not. This leads to poor quality sequence data. The use of a single-gel track for all four dideoxy reactions means that this problem is less acute in automated sequencing. Nevertheless, it is desirable to sequence a piece of DNA several times, and on both strands, to eliminate errors caused by technical problems. It should be noted that long runs of

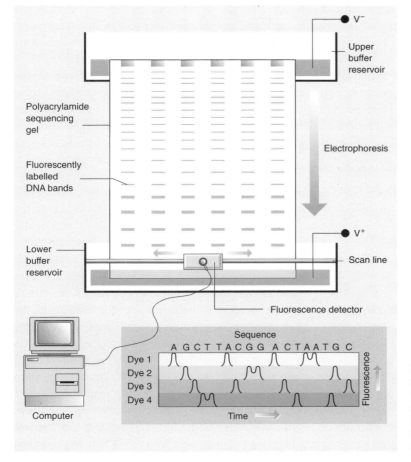

Figure 5.6 Block diagram of an automated DNA sequencer and idealized representation of the correspondence between fluorescence in a single electrophoresis lane and nucleotide sequence.

the same nucleotide or a high G + C content can cause compression of the bands on a gel, necessitating manual reading of the data, even with an automated system. Note also that multiple, tandem short-repeats, which are common in the DNA of higher eukaryotes, can reduce the fidelity of DNA copying, particularly with Taq DNA polymerase. The second advantage of automated DNA sequencers is that the output from them is in machine-readable form. This eliminates the errors that arise when DNA sequences are read and transcribed manually.

The sequencing rate

The theoretical sequencing rate is easy to calculate. It is the number of sets of sequencing reactions that can be loaded on each gel, times the number of bases read from each sample, times the number of gels that can be run at once, times the number of days this can be carried out per year. For a 24-channel fluorescent sequencer this, in theory, is about 2.7 million bases (Mb) per machine per year. Similar calculations can be made for manual sequencing. Until recently (Barrell 1991) the best that had been achieved was about 25 kb per person per year. For this reason a number of groups attempted to improve the sequencing throughput, e.g. by multiplexing (Church & Kieffer-Higgins 1988; Cherry *et al.* 1994). Recently, however, Fleischmann *et al.* (1995) were able to generate over 500 kb of *completed* sequence per automated machine per year and the throughput per person year was 450 kb. With such productivity the need for changes to the basic dideoxy sequencing method are less pressing than they were.

Strategies for using Sanger sequencing

Two basic genome sequencing strategies have evolved: complete genome sequencing and cDNA sequencing. The feasibility of the former strategy has been shown by the assembly of the entire sequence of the genome of the yeast *Saccharomyces cerevisiae* (Goffeau *et al.* 1996). By definition, total genome sequencing generates detailed and complete data. However, it demands accuracy and is very labour intensive. Since the maximum length of DNA that can be sequenced in a single pass is 300–500 bases the sequencing of an entire chromosome or genome means that many thousands of short overlapping stretches need to be analysed. A potential way of reducing the workload is to use the directed or primer walking approach (Kieleczawa *et al.* 1992). This involves sequences being continually extended through the design of oligonucleotides required to prime further sequencing reactions. Originally it was thought that this would require large amounts of unique primers. However, Kieleczawa *et al.* (1992) have shown that hexamers are able to assemble into long, unique primers upon annealing to adjacent positions on the template. Thus, for example, from a library of 4096 hexamers all unique 18mers could be created and Ghiso *et al.* (1993) have suggested that an optimized subset of 1200 hexamers could be sufficient for many studies. It remains to be seen if this modification will be adopted by others.

To date, most whole genome sequencing projects have taken place in two stages: construction of a physical map and sequencing of large numbers of genome fragments. The physical map is required so that the sequenced fragments can be reassembled. It should be noted that with this methodology the most difficult and time-consuming task is *not* the sequencing but the construction of the physical map. For this reason Fleischmann *et al.* (1995) developed a shotgun sequencing strategy, which dispenses entirely with the mapping stage. With this strategy, a single, random DNA fragment library is constructed. The ends of a sufficient number of randomly selected fragments are sequenced and assembled to produce a complete genome. This strategy has been employed successfully in the sequencing of a number of bacterial genomes, e.g. Fleischmann *et al.* (1995), Fraser *et al.* (1995) and Bult *et al.* (1996). Venter *et al.* (1996) have suggested how it might be extended to larger genomes. A comparison of these two strategies is shown later in Fig. 5.10.

One problem with all genomic sequencing projects is the vast amount of data generated. As a rough guide, 5–10 kb of data are required to assemble 1 kb of confirmed sequence. The effort involved needs to be assessed in terms of the information derived from it. In prokaryotes and lower eukaryotes the gene density is very high (~ 900 genes per Mb) because introns are rare or absent and there is relatively little repetitive DNA. This is reduced to about 200 genes per Mb of DNA in the nematode (Wilson *et al.* 1994) and about 600 genes per Mb in yeast (Oliver *et al.* 1992). By contrast, a megabase of human DNA is expected to contain only 10 or 20 genes on average (Barrell 1991), although the gene density shows great regional variation.

An alternative sequencing strategy for higher eukaryotes is to analyse only cDNA. For example, in sequencing a highly spliced gene scattered in many exons over a megabase of human DNA, it would be much simpler to sequence the same coding region in 2 kb of cDNA rather than a megabase of genomic DNA. However, if the cDNA sequence is established first there is little incentive to sequence the genomic DNA at a later date! If one has only the genomic sequence, then the cDNA sequence is needed to identify unambiguously the introns and exons. A variant of the cDNA sequencing approach has been adopted by a number of groups associated with the human genome project. In this, clones are selected at random from a cDNA library and a small amount of sequence determined for each, e.g. 200–300 nucleotides, a length that can be obtained in a single sequencing run. This information is incomplete as the full sequence of a cDNA usually comprises 1000–10 000 nucleotides. Furthermore, a low rate of errors in the sequence is not important, for the sequence acts merely as a 'signature'. These signatures are used to scan databases to determine whether they correspond to known genes. If so, no further work is necessary. If not, the entire cDNA sequence can be sequenced. Okubo *et al*. (1992) and Adams *et al*. (1995) are using this approach with many different human cells and tissues in order to construct a 'body map of expressed human genes'.

While cDNA sequencing reduces the workload, e.g. expressed genes constitute only 5% of the human genome, it still is technically demanding. The problems with this approach are fourfold. First, control elements such as enhancers and promoters as well as splice sites would not be sequenced. Second, many genes are expressed at very low levels or for very short periods and so may not be represented in cDNA libraries. Third, there is some debate as to what actually constitutes a gene and therefore there is no agreement as to how many genes there are in each genome (for discussion see Fields *et al*. 1994; Bird 1995). Finally, much of the DNA not sequenced, so-called 'junk' DNA, could have functions that are as yet unknown. This is suggested from the high degree of cross-species conservation such as in the region of the T-cell receptor gene (Koop 1995). Furthermore, when Mantegna *et al*. (1994) extended the Zipf approach to analysing linguistic texts to the statistical study of DNA base pair sequences they found that the non-coding regions are more similar to natural languages than the coding regions. By using an approach that quantifies the 'redundancy' of a linguistic text in terms of a measurable entropy function they were able to show that non-coding regions in eukaryotes display a smaller entropy and larger redundancy than coding regions. This supports the possibility that non-coding regions of DNA may carry biological information.

Krishnan *et al*. (1995) recently have suggested a third sequencing strategy called *feature mapping*, which is used to help guide the choice of regions to be subjected to detailed analyses.

Feature mapping

In feature mapping, large DNA fragments are cloned into a transposon-based cosmid vector designed for generating nested deletions by *in vivo* transposition and simple bacteriological selection. These deletions place primer sites throughout the DNA of interest at locations that are easily determined by plasmid size (Fig. 5.7). In this way, Krishnan *et al*. (1995) generated 70 informative deletions of a 35 kb human DNA fragment. DNA adjacent to the deletion end points was sequenced and constituted the foundation of a feature map by: (i) identifying putative exons and positions of *Alu* elements; (ii) determining the span of a gene sequence; and (iii) localizing evolutionarily conserved sequences.

Conventional genomic sequencing

The phage, cosmid or YAC clones used in the preparation of physical maps are too large to be sequenced directly. Consequently, clones are subjected to random sub-cloning prior to sequencing. Early in the project this results in rapid accumulation of sequence data but later is subjected to the law of diminishing returns. At a point where approximately 95% of a sequence has been determined, a switch is made to a directed approach to obtain the remainder of the sequence. Note that for this strategy cosmids are the vector of choice because a large amount of DNA can be purified (cf. YACs) and the ratio of insert : vector sequence is high (9 : 1) (cf. phage vectors). The detailed procedure described here is that of Wilson *et al.* (1992). All genomic sequencing groups follow a similar strategy although the detailed tactics may vary. In particular, many groups are moving to the use of cycle sequencing (p. 100).

The first step is to fragment the cloned DNA in a random manner and sonication has proved to be the most effective method. After sonication, fragments in the size range 0.8–1.6 kb are selected by preparative agarose gel electrophoresis and sub-

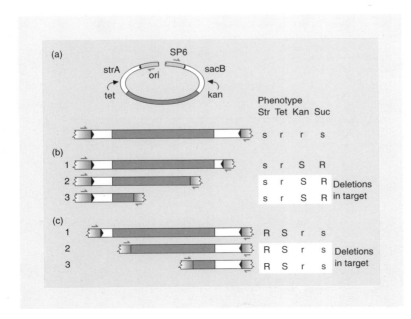

Figure 5.7 Schematic representation of hypothetical nested deletions obtained using a transposon-based cosmid vector. (a) Depiction of a hypothetical DNA clone. The green portion represents cloned DNA, the filled arrowheads the transposon sequences and the green half-arrows represent locations of primer binding sites. (b) Deletions extending into various sites within the cloned fragment resulting from selection for resistance to sucrose, caused by loss of sacB, and tetracycline. (c) Deletions resulting from selection for resistance to streptomycin, caused by loss of strA, and kanamycin. R/r indicates resistance and S/s indicates sensitivity. (Redrawn from Krishnan *et al.* 1995, by permission of Oxford University Press.)

cloned in a suitable M13-based sequencing vector. Smaller frag-ments are less useful for sequencing, whereas larger fragments can undergo deletion during propagation in M13. The next step is to sequence random sub-clones and this is achieved using the M13 universal primer. For accuracy, the same sequence needs to be obtained from 5–6 different clones and should be obtained for both strands. The number of different, randomly selected sub-clones that need to be sequenced to cover 95% of the original cosmid clone depends on the size of that primary clone and can be determined using the statistical profiles of Bankier & Barrell (1983). After DNA sequence analysis of the appropriate number of random sub-clones, sequence contigs are assembled using appropriate computer programs. Gaps in the contigs have to be closed and this is carried out by synthesizing PCR primers homol-ogous to the ends of the contigs and using them to isolate the missing sequence (Fig. 5.8). This primer-directed 'walking strat-egy' is continued until all the gaps have been closed.

The report of Wilson *et al.* (1992) provides detailed informa-tion on the logistics of sequencing a large piece of DNA, in this case 96 kb of mouse DNA. Initially they used radioisotopic methods and a relatively high throughput was maintained using robotics to perform DNA sequencing reactions and to load gels. Up to 24 sub-clones could be processed per person per day and another full day was required to process the data generated. The maximum throughput obtained was 96 sub-clones per person per week. To reduce the burden of manual data entry from autorad-iographs they switched to automated sequencing instruments

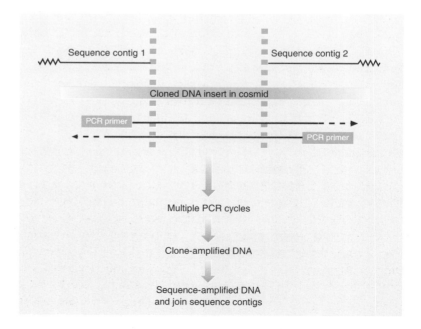

Figure 5.8 Linking DNA sequence contigs by walking.

based on fluorescent dye chemistry. Under optimal conditions it was possible to process up to 120 sub-clones per instrument per week. The magnitude of the task can be appreciated when it is realized that Wilson *et al.* (1992) sequenced 2399 sub-clones in order to generate their 96 kb of mouse DNA sequence! A significant fraction (27.5%) of these sub-clones did not generate useful sequence because they either lacked an insert or a primer site or could not be produced in sufficient amounts for sequence analysis. In addition, using the automated analyser, instrument failures and downtime are significant (15–20%). It remains to be seen whether some of the more recent improvement (p. 99) to the basic Sanger sequencing method can significantly improve the throughput in large-scale sequencing procedures of this kind.

Sequence accuracy

A number of the sub-clones generated in the above sequencing procedure will contain fragments of the cosmid vector. Some groups (e.g. Edwards *et al.* 1990) have described the identification and exclusion of such sub-clones using plaque hybridization. Wilson *et al.* (1992) chose not to direct any efforts into pre-screening but to process the relatively small number (12%) of sub-clones containing regions of the cosmid vector. Not only did it eliminate tedious plaque hybridization but it allowed error analysis by comparison of raw sequence data with known vector sequences. Sequences that were read beyond 400 bp contained an average of 3.2% error, while those less than 400 bp had 2.8% error. At least one-third of the errors were due to ambiguities in sequence reading. In those sequences longer than 400 bp that were read, most errors occurred late in the sequence and often were present as extra bases in a run of two or more of the same nucleotide. The remainder of the errors were due to secondary structure in the template DNA. However, because the complete sequence was analysed with an average 5.9-fold redundancy and most of it on both strands, the final error frequency is estimated to be less than 0.1%. By comparison, 35 different European laboratories were engaged in sequencing the *S. cerevisiae* genome with the attendant possibility of a very high error frequency. However, by using a DNA coordinator who implemented quality control procedures (Table 5.1), the overall sequence accuracy for yeast chromosome XI (666 448 bp) was estimated to be 99.97% (Dujon *et al.* 1994), i.e. similar to that noted above. Lipshutz *et al.* (1994) have described a software program for estimating DNA sequence confidence. Fabret *et al.* (1995) have analysed the errors in finished sequences. They took advantage of the fact that the surfactin operon of *Bacillus subtilis* had been sequenced by three independent groups. This enabled the *actual* error rate to be calculated. It was found to range from 0.02 to 0.27%, the different error rates being ascribed to the detailed sequencing tactics used. Other studies of DNA sequencing accuracy have been summarized by Yager *et al.* (1997). In general, a sequence read of > 350 nucleotides at 99% accuracy can be expected using current ultrathin slab gel technology. Reading lengths in excess of 1000 nucleotides have been reported (Noolandi *et al.* 1993; Voss *et al.* 1995).

Genome sequencing without prior mapping

The paradigm for this mode of genomic sequencing is the determination of the 1 830 137 bp sequence of the bacterium *Haemophilus influenzae* (Fleischmann *et al.* 1995). The starting point was the preparation of genomic DNA, which then was mechanically sheared and size fractionated. Fragments between 1.6 and 2.0 kb in size were selected, this narrow range being chosen to minimize variation in growth of clones. In addition, with a maximum size of only 2.0 kb the number of complete genes on a DNA fragment is minimized thereby reducing the chance of their loss through expression of deleterious gene products. The selected fragments were ligated to a sequencing vector and again size fractionated to minimize contamination from double-insert chimaeras or free vector. Finally, all cloning was undertaken in host cells deficient in all recombination and restriction functions to prevent deletion and rearrangement of inserts.

In all, 28 643 sequencing reactions were undertaken on inserts and this took eight individuals using 14 automated sequencers a total of 3 months. The collected sequence data was then assembled into contigs using sophisticated algorithms. Starting with an initial fragment a second candidate fragment was chosen with the best overlap based on oligonucleotide content. The contig was extended by the fragment only if strict match criteria were met.

Table 5.1 Quality control of sequence data from 35 laboratories engaged in sequencing yeast chromosome XI. (Data from Dujon *et al.* 1994 with permission from *Nature*, © Macmillan Magazines Ltd)

Method of verification	Total number of fragments	Total bp verified	Error % detected
Original overlap between cosmids	28	63 424	0.02
Resequencing of selected segments (3–5 kb long)	21	72 270	0.03
Resequencing of random segments (~ 300 bp long)	71	18 778	0.05
Resequencing of suspected segments (~ 300 bp long) from designed oligonucleotide pairs	60	17 035	0.03
Total	180	171 507	
Average error rate			0.03

These match criteria included the minimum length of overlap, the maximum length of an unmatched end and the minimum percentage match. A total of 24 304 sequences were deemed useful and were assembled into 140 contigs after 30 h of central processing unit time. The 140 contigs were separated by 42 physical gaps (no template DNA for the region) and 98 sequence gaps (template available for gap closure). The sequence gaps were closed by PCR walking as described previously (see Fig. 5.8).

A number of techniques were used to close the physical gaps; for example, oligonucleotide primers were designed and synthesized from the end of each contig. These primers were used in hybridization reactions based on the premise that labelled oligonucleotides homologous to the ends of adjacent contigs should hybridize to common DNA restriction fragments (Fig. 5.9a). Links also were made by searching each contig end against a peptide database. If the ends of two contigs matched the same database sequence, then the two contigs were tentatively assumed to be adjacent (Fig. 5.9b). Finally, two λ libraries were constructed from genomic *H. influenzae* DNA and were probed with the oligonucleotides designed from the ends of each contig. Positive plaques then were used to prepare templates and the sequence was determined from each end of the λ clone insert. These sequence fragments were searched against a database of all contigs and two contigs that matched the sequence from the opposite ends of the same λ clone were ordered (Fig. 5.9c). The λ clone then provided the template for closure of the sequence gap.

The λ clones were particularly useful for solving repeat structures. All repeat structures identified in the genome were small enough to be spanned by a single clone from the random insert library, except for the six ribosomal RNA (rRNA) operons and one repeat (two copies) that was 5340 bp in length. The ability to distinguish and assemble the six rRNA operons was a test of the potential effectiveness of the sequencing strategy used if it were to be applied to the sequencing of a more complex genome that contains a significant number of repeat regions. The high degree of sequence similarity and the length of the six operons caused the assembly process to cluster all the underlying sequences into a few indistinguishable contigs. To determine the correct placement of the operons in the sequence, unique sequences were identified at the ends of the 5S RNA genes. Oligonucleotide primers were designed from these six flanking regions and used to probe the two λ libraries. For five of the six rRNA operons, at least one positive plaque was identified that completely spanned the rRNA operon and contained uniquely identifying flanking sequence at the 16S and 5S ends. These plaques provided the template for obtaining the sequence for these RNA operons.

As far as accuracy is concerned, it is estimated that there is one error per 5–10 000 bases for the *H. influenzae* sequence and one error per 10 000 bases for *Mycoplasma genitalium* (Fraser *et al.* 1995).

The sequence-tagged connectors approach

There are a number of problems associated with conventional genomic sequencing. First, as already mentioned above, there is a need for detailed physical maps whose construction is both tedious and lengthy. Second, there is a dependence on YACs. These show structural instability of inserts resulting in deletions or rearrangements of portions of the cloned DNA. Problems also are presented by chimaeras in which two non-adjacent portions of the genome end up in one clone. Finally, the genomes of higher eukaryotes contain tandem arrays of DNA units with high sequence similarity. These pose problems for high resolution mapping and sequencing when the size of the clone insert is less than that of the tandem array because the landmarks are similar. Consider, for example, a human genome sequence with five tandem 21 kb arrays. This is larger (105 kb) than the maximum insert size of a cosmid (40 kb) which will be used for sub-cloning.

Because of the problems noted above, Venter *et al.* (1996) have *proposed* an alternative approach. This builds on the success, detailed in the previous section, with whole genome sequencing in the absence of a physical map. Rather than using YACs and cosmids it makes use of BACs. These can accept DNA inserts up to 350 kb in length and appear to be much more stable than YACs. Also, they appear to be good substrates for shotgun sequencing. The basic principle of the method is shown in Fig. 5.10. A BAC library is prepared which has an average insert size of 150 kb and a 15-fold coverage of the genome in question. The individual clones making up the library are arrayed in microtitre wells for ease of manipulation. Starting at the vector-insert points, both ends of each BAC clone are then sequenced to generate 500 bases from each end. These BAC end-sequences will be scattered approximately every 5 kb across the genome and make up 10% of the sequence. These 'sequence-tagged connectors', or STCs, will allow any one BAC clone to be connected to about 30 others.

Each BAC clone would be fingerprinted using one restriction enzyme to provide the insert size and detect artefactual clones by comparing the fingerprints with those of overlapping clones. A seed BAC of interest is sequenced and checked against the database of STCs to identify the 30 or so overlapping BAC clones. The two BAC clones showing internal consistency among the fingerprints and minimal overlap at either end then would be sequenced. In this way the entire human genome could be sequenced with just over 20 000 BAC clones. It should be stressed that this method has not yet been reduced to practice and not everyone is convinced of its practicability (Little 1996). The major criticism of the proposed method is the concept of simply assembling short-insert sequences into finished sequence for complex genomes. The problem is not so much the tandem repeats, as

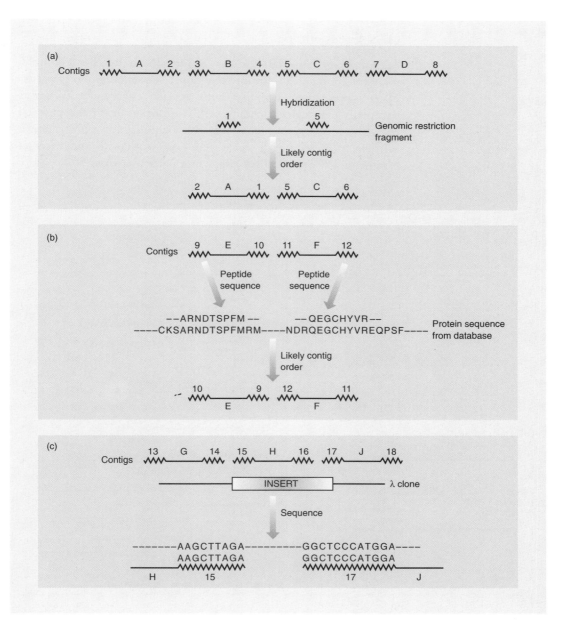

Figure 5.9 The three methods for closing the physical gaps in sequencing by the method of Fleischmann *et al.* (1995). In each case the sequences at the ends of contigs are shown as wavy lines and individual sequences are given separate numbers. Contigs are denoted by large capital letters. In (b) the individual amino acids are represented by the standard single letter code. See text for a detailed description of each method.

discussed above, but the dispersed repeats. For example, 20 kb of human DNA can contain over 30 Alu elements, often very similar in sequence, in tandem or inverted arrays (see p. 41). If an autoassembler program were to be used to generate the finished sequence, then there would be a high likelihood of spurious inversions and deletions.

cDNA sequencing

As noted earlier (p. 103), cDNA sequencing represents a complementary approach to complete genome sequencing and is particularly appropriate for use with higher eukaryotes. Whereas complete genome sequencing yields information on the physical structure of the genome, cDNA sequencing can give information on gene expression, e.g. what genes are expressed, and to what extent, in any given cell or tissue at a particular time.

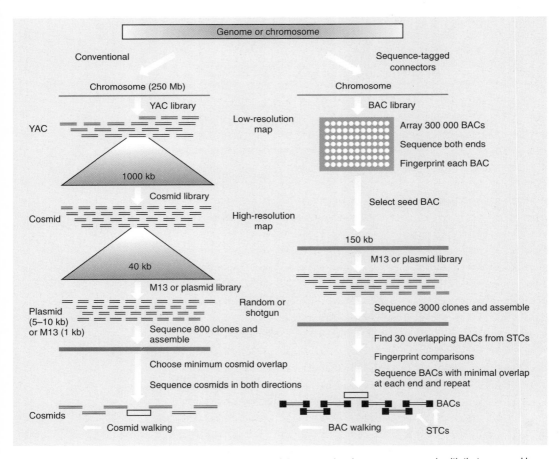

Figure 5.10 Comparison of the conventional sequence approach with that proposed by Venter *et al.* (1996). See text pp. 110–11 for full details. BAC, bacterial artificial chromosone; STC, sequence-tagged connector. (Redrawn from Venter *et al.* 1996 with permission from *Nature*, © Macmillan Magazines Ltd.)

There are several different strategies for constructing a cDNA library that can greatly influence the available set of information. First, a cDNA library may be prepared from a selected cell or organ so the library constituents reflect the physiology of that cell or organ (Weinstock *et al.* 1994). Alternatively, combining several cDNA sub-libraries from various tissues maximizes the number of gene transcripts. Second, the library may be designed so as to represent faithfully the abundance of gene transcripts in the original mRNA population or, alternatively, to represent a collection of non-overlapping transcripts. The latter type, a *single book* library, is used for surveying as many gene transcripts as possible. The generation of a single book library requires the use of a subtractive cDNA cloning procedure, preferably one that needs small amounts of starting mRNA (Travis & Sutcliffe 1988) or a cloning procedure which specifically generates a nearly equal representation of cDNAs (Ko 1990). Without the use of such procedures, low abundance mRNAs are unlikely to be cloned as cDNAs. Finally, the library may be designed to carry full-length inserts. Alternatively, it may consist of partial cDNAs covering a directed portion of each mRNA, such as the 3' or 5' ends. Composition of the latter libraries can be proportional to the composition of the mRNA population, in contrast to the full-size library. Sequencing from the 3' end of cDNA clones generates mainly 3' untranslated sequence which is helpful in discriminating between homologous gene family members but gives little information about gene function. It also results in repeated sequencing of uninformative poly(dA) stretches despite the use of oligo(dT) primers. Although 5' sequencing can overcome these limitations, varying clone lengths mean that sequence information may not be generated from the same position in homologous clones. Thus true identity may be missed.

Using a random selection approach, Adams *et al.* (1991) found an unacceptably large number of highly represented clones in their cDNA libraries. Over 30% of clones from human hippocampal DNA were uninformative. Nevertheless, from 600 clones sequenced, 337 represented new genes. In an extension of this work, Adams *et al.* (1992) sequenced an additional 2672 independent clones and identified 2375 new sequences. To eliminate highly represented cDNAs, Khan *et al.* (1992) pre-screened their library with labelled total brain cDNA and an excess of cold genomic DNA. Plaques that were not labelled were selected and gave a sevenfold increase in the number of new sequences identified. Of 1024 human brain cDNAs, over 900 represented new human genes. Okubo *et al.* (1992) used as the basis of their sequencing a 3'-directed cDNA library (Fig. 5.11) from a liver cell line which is a non-biased representation of the mRNA population. In total, 982 random cDNA clones were sequenced and, because of the

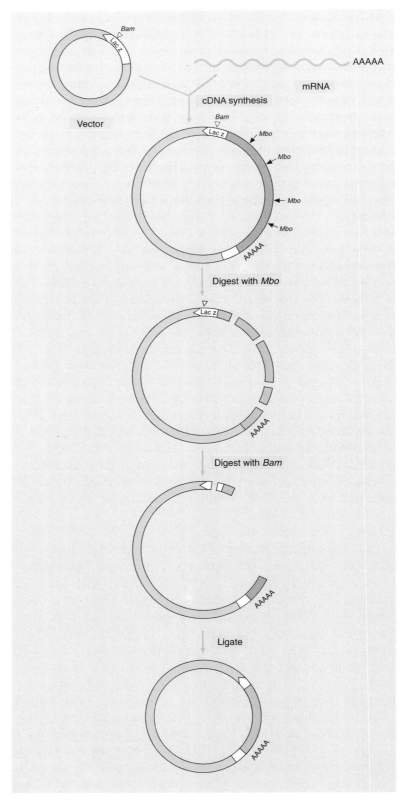

Figure 5.11 The method used to clone the 3′ end of a cDNA. The sequencing vector DNA is methylated prior to cloning the cDNA to make it resistant to endonuclease *Mbo*I. This can be achieved by *in vitro* methylation or by preparing the vector from a *dam* host. The mRNA is converted to cDNA using the Okayama and Berg procedure (see Old & Primrose 1994). The clones carrying the cDNA are digested with *Mbo*I to leave only the 5′ and 3′ ends of the cDNA. The 5′ end then is removed by digestion with *Bam*HI. Note that the overhangs produced by endonucleases *Mbo*I and *Bam*HI are identical.

random selection of clones, several identical sequences emerged. The frequency distribution of these is shown in Fig. 5.12. Altogether, among the 982 clones that produced informative sequence data, 215 clones represented 91 known genes and 767 clones represented 550 novel genes. Adams *et al.* (1993a,b) have reported the sequencing of yet another 5000 cDNAs and once again many interesting genes have been identified. Whereas Adams *et al.* (1993a,b) concentrated on brain cDNAs, Okubo *et al.* (1992) studied liver cDNAs. Comparison of the sequence profiles from these two tissues indicates that structural and regulatory proteins are more abundant in the brain but secreted proteins and polypeptides involved in protein synthesis predominate in the liver.

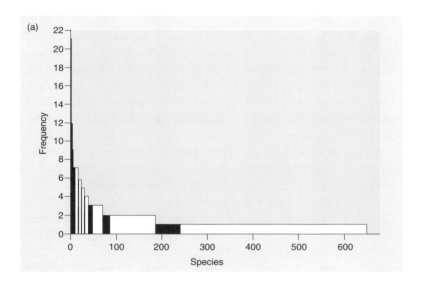

Frequency (in 982)	Representative clone	Insert size (nt)	Gene
22	a29	523	Elongation factor in 1α
21	c12a04	132	Serum albumin
12	a26	188	Translationally controlled tumour protein
9	c12a01	261	Unknown
8	a31	369	α-1-antitrypsin
7	a23	183	Rat ribosomal protein L21
7	c12b03	25	Ribosomal protein L31
7	c12b06	88	Rat ribosomal protein S19
7	c13c04	174	Mouse ribosomal protein L3
7	d2g03	101	Unknown
6	a-30	332	Ferritin light chain
6	a-31	156	Unknown
6	a20	383	Unknown
6	d0g12	107	Unknown
6	hm05a02	373	Unknown
6	s304	120	Unknown
5	a29	182	Ribosomal protein S16
5	a32	190	Apolipoprotein A11
5	hm02g10	163	Unknown
5	s406	364	Unknown

Figure 5.12 Expression profiles of genes in a human liver cell line. The frequencies of the appearance of each cDNA species in 982 randomly sequenced clones are shown in (a). Species are arranged in the order of their frequencies of appearance. Solid and open bars represent the novel and identified sequences respectively. (b) This table shows some properties of the most abundant 20 species. (Redrawn with permission from Okubo *et al.* 1992, courtesy of *Nature Genetics*.)

In the most extensive cDNA analysis to date, Adams *et al.* (1995) examined 174 thousand ESTs generated from 248 primary and 52 subtracted cDNA libraries constructed from 37 distinct organs and tissues. These ESTs were combined with another 118 thousand in an EST database. Taken together they represent 83 million nucleotides of information. Because of the inherent redundancy of this cDNA sampling it is possible to build assemblies of ESTs, essentially treating the expressed portion of the genome as a shotgun sequence assembly project. In many cases these assemblies, termed tentative human consensus sequences (THCs), represent full-length transcripts and over 29 thousand were recorded by Adams *et al.* (1995).

In Okubo *et al.* (1992), their work with 982 random cDNA clones generated over 270 kb of sequence data. As noted earlier (p. 107) the accuracy of this sequence data is important. If the sequence data are to be used for predicting translation products, sequencing errors need to be minimized by multiple sequencing efforts. On the other hand, if the sequence data are used only for identifying mRNA species, handling a large number of samples is more important than the accuracy of the sequence data. In all the studies to date, the number of mRNA species identified has been the key criterion and single pass sequencing has been used.

A logical extension of single pass sequencing of cDNAs is determination of the precise chromosomal location of each of the corresponding genes. While cDNAs can be used directly for FISH mapping to metaphase chromosomes, this can be technically difficult. Also, while regional mapping information is generated, no physical genomic clone corresponding to the cDNA is produced. However, mapping cDNAs to genomic cosmids or YACs and then using the latter as FISH probes provides both regional mapping data and a cloned piece of genomic DNA from which the cDNA is derived. Polymeropoulos *et al.* (1993) have mapped more than 300 human brain ESTs. Surprisingly, their distribution did not correlate with the cytological length of the chromosomes. For example, chromosomes 2 and 19 contain similar numbers of genes despite their large difference in size, whereas chromosomes 13 and 18 were relatively under-represented. The latter observation may explain why the only viable human trisomies are of chromosomes 13, 18 and 21.

In the vast majority of cases, the demonstration of sequence identity between a newly obtained EST and a previously described gene will be correctly interpreted as evidence for the expression of that specific gene in the tissue from which the EST was derived. However, Tsai *et al.* (1994) have shown that this is not always the case. They showed that a gene called *TCTE1* was not expressed in tissues other than the male reproductive track, whereas Adams *et al.* (1992, 1993a,b) reported putative *TCTE1* ESTs in cDNA clones from human brain libraries. However, detailed analysis showed that EST-defined clones with homology to *TCTE1* sequences are derived from a transcription unit present on the opposite strand of *TCTE1*.

The use of cDNA sequencing is not restricted to human DNA. Sequencing projects are underway in *Caenorhabditis* (McCombie *et al.* 1992a; Waterston *et al.* 1992), *Arabidopsis* (Cooke *et al.* 1996), the malarial parasite *Plasmodium falciparum* (Chakrabarti *et al.* 1994), the agent of toxoplasmosis *Toxoplasma gondii* (Wan *et al.* 1995), rice (Kurata *et al.* 1994; Sasaki *et al.* 1994), *Drosophila* and the mouse (Boguski *et al.* 1993; Sikela & Auffray 1993). Data now are being generated for ESTs at such a rate that by October 1997 ESTs were available for over half of the estimated 70 000– 100 000 human genes. A special database (dbEST) has been constructed specifically to meet the unique informatics challenges posed by this wealth of data.

Alternative sequencing techniques

Because of the workload associated with generating long stretches (> 100 kb) of sequence data, much of it technically demanding but extremely boring, attempts have been made to automate as much of it as possible. Automated sequencing machines have been available for a number of years but most of the labour is required in earlier steps, e.g. to select clones, make DNA preparations, carry out the sequencing chemistry and pour and load sequencing gels. Equipment that will automate some of these steps is beginning to become commercially available.

Many attempts are being made to miniaturize the current slab gel technology. Successful separation of DNA ladder bands beyond 430 bases has been achieved in 10 min on a silicon chip that has been etched to contain an array of channels with a 3.5 cm separation length (Wooley & Mathies 1995). An alternative way to speed up DNA sequencing would be to replace the conventional slab gel electrophoresis with capillary electrophoresis. However, the routine preparation of homogeneous gels in capillaries is not a simple task because of gel shrinkage and the formation of bubbles inside the capillary. New sieving polymers have been developed which may obviate these problems (Quesada 1997).

An alternative approach to increasing throughput is to develop new sequencing methodology. Over the years a number of novel techniques have been proposed. Despite generating a lot of initial excitement most have sunk without trace. Only two still are considered credible: sequencing by hybridization (see next section) and scanning tunnelling microscopy (STM). In STM, a very fine tip is kept extremely close to the object, in this case DNA, by a control system based on the detection of a minute current induced by tunnel effect between the tip and the DNA. In the related atomic force microscopy (AFM) control is based on measurement of the Van der Waals interaction force between the probe and the sample. Either way, the tip is moved to scan across the object, its vertical displacements being precisely measured and stored thereby generating a picture of a sample surface. Reasonable pictures have been taken of single-stranded DNA (Dunlapp & Bustamante 1989) and double-stranded DNA (Driscoll *et al.* 1990) but it remains to be seen if individual bases in DNA can be recognized. If they can, then the technique has the potential to generate 1 Mb of sequence data in less than a day.

The continued refinement of two recent methods for producing gas-phase ions, electrospray ionization and matrix-assisted laser desorption ionization, has resulted in new techniques for the rapid characterization of oligonucleotides by mass spectrometry. However, it is considered unlikely that mass spectrometry will replace existing methods for sequencing oligos much longer than 100mers (Limbach *et al.* 1995).

Sequencing by hybridization

The principle of sequencing by hybridization can best be explained by starting with a simple example. Consider the tetranucleotide CTCA, whose complementary strand is TGAG, and a matrix of the whole set of $4^3 = 64$ trinucleotides. This tetranucleotide will specifically hybridize only with complementary trinucleotides TGA and GAG revealing the presence of these blocks in the complementary sequence. From this the sequence TGAG can be reconstructed. If instead of using trinucleotides, $4^8 = 65\,536$ octa-nucleotides were used, it should be possible to sequence DNA fragments up to 200 bases long (Bains & Smith 1988; Lysov *et al.* 1988; Southern 1988; Drmanac *et al.* 1989). The length of the target that can be analysed is approximately equal to the square root of the number of oligonucleotides in the array (Southern *et al.* 1992). Two different experimental configurations have been developed for the hybridization reaction. Either the target sequence may be immobilized and oligonucleotides labelled (Fig. 5.13) or the oligonucleotides may be immobilized and the target sequence labelled. Each method has advantages over the other for particular applications. It is an advantage to label the oligonucleotides to analyse a large number of target sequences for fingerprinting. On the other hand, for applications that require large numbers of oligonucleotides of different sequence, it is advantageous to immobilize oligonucleotides and use the target sequence as the labelled probe.

Drmanac *et al.* (1992) have developed a protocol based on the first method and which has the theoretical capability of determining up to 100 Mb per year. Indeed, 20 replica filters containing 0.5 million clones, representing 100 Mb, can be prepared by one individual in one month. Typical results that can be obtained are shown in Fig. 5.14. More recently, Drmanac *et al.* (1993) successfully used several hundred 8–10mer probes to sequence three 343 base fragments of DNA from phage M13. Meier-Ewert *et al.* (1993) have described how this method can be used as an alternative to gel-based methods partially to sequence cDNA clones (see p. 103). Southern *et al.* (1992) have developed model systems based on the second configuration (Fig. 5.15) and demonstrated its feasibility. They developed technology for making complete sets of oligonucleotides of defined length and covalently attaching them to the surface of a glass plate by synthesizing them *in situ*. A device carrying all octapurine sequences was used to explore factors affecting molecular hybridization of the tethered oligonucleotides, to develop computer-aided methods for analysing the data, and to test the feasibility of using the method for sequence analysis.

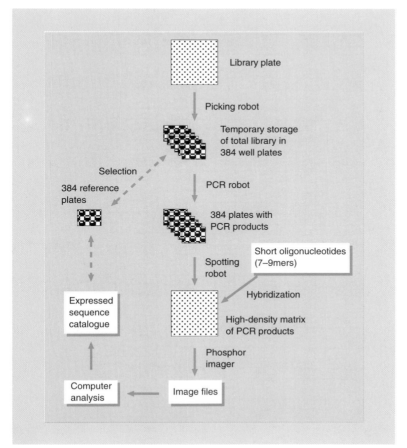

Figure 5.13 Outline of scheme for sequencing by hybridization where labelled probes are applied to cloned DNA immobilized on membranes. (Redrawn from Meier-Ewert *et al.* 1993 with permission from *Nature*, © Macmillan Magazines Ltd.)

Hybridization to sequences on microchips

Recently, there has been a marriage between light-directed synthetic (photolithography) methods routinely used in the semiconductor industry and standard oligonucleotide synthesis using combinatorial methods (Fodor *et al.* 1993; Pease *et al.* 1994; Southern 1996). The result is that, in about one day, a series of microchips can be prepared containing potentially hundreds of thousands of oligonucleotides of predetermined sequence. The process is deceptively simple. In the first step a mercury lamp is

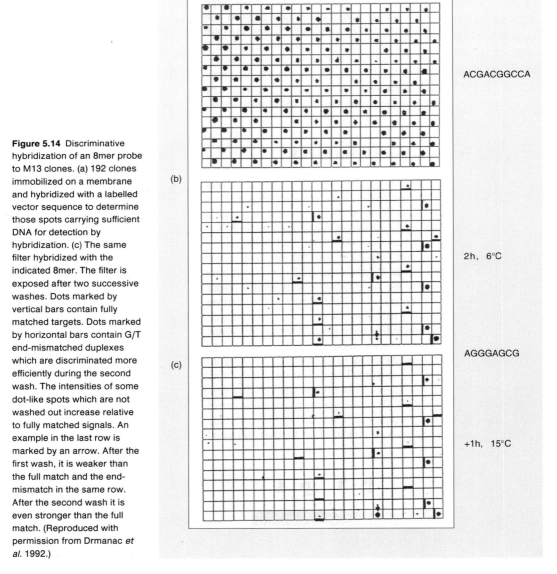

(a)

ACGACGGCCA

(b)

2h, 6°C

AGGGAGCG

(c)

+1h, 15°C

Figure 5.14 Discriminative hybridization of an 8mer probe to M13 clones. (a) 192 clones immobilized on a membrane and hybridized with a labelled vector sequence to determine those spots carrying sufficient DNA for detection by hybridization. (c) The same filter hybridized with the indicated 8mer. The filter is exposed after two successive washes. Dots marked by vertical bars contain fully matched targets. Dots marked by horizontal bars contain G/T end-mismatched duplexes which are discriminated more efficiently during the second wash. The intensities of some dot-like spots which are not washed out increase relative to fully matched signals. An example in the last row is marked by an arrow. After the first wash, it is weaker than the full match and the end-mismatch in the same row. After the second wash it is even stronger than the full match. (Reproduced with permission from Drmanac *et al.* 1992.)

120

shone through a standard computer-industry photolithographic mask onto the synthesis surface. This activates specific areas for chemical coupling with a nucleoside which itself contains a 5′ protecting group. Further exposure to light removes this group, leaving a 5′-hydroxyl group capable of reacting with another nucleoside in the subsequent cycle. The choice of which nucleoside to activate is thus controlled by the composition of the mask. Successive rounds of deprotection and chemistry can result in an exponential increase in oligonucleotide complexity on a chip for a linear number of steps. For example, it requires only $4 \times 15 = 60$ cycles to synthesize a complete set of ~ 1 billion different 15mers. The space occupied by each specific oligonucleotide is termed a feature and may house at least 1 million identical molecules (Fig. 5.16). A standard 1.6 cm square chip can house more than 1 million different probes with a spacing of 1 μm, cf. the computer industry which routinely achieves resolution at 0.3 μm!

To date, microchips have not been used for sequencing *de novo* but have been used for *resequencing* a genome, that of human mitochondrial DNA, and for detecting polymorphisms (Chee *et al.*

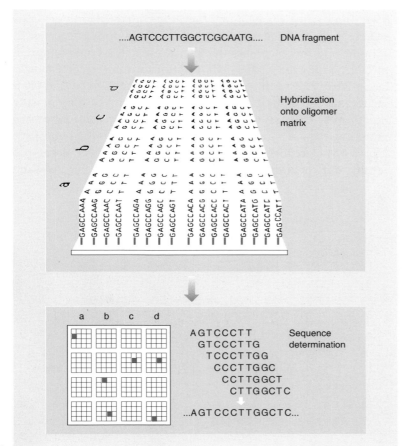

Figure 5.15 Schematic representation of sequencing by hybridization where the sequence being examined is labelled and hybridized with immobilized oligomers. (Redrawn with permission from Hoheisel 1994.)

1996). For this purpose a 4L tiled array is set up in which L corresponds to the length of the sequence to be analysed. The sequence is probed with a series of oligomers of length P which match exactly the target sequence except for one position which is systematically substituted with each of the four bases A, T, G or C. In the example shown in Fig. 5.17, a tiled array of 15mers varied at position 7 from the 3′ end is used. This is known as a $P^{15,7}$ array. To use such an array, the DNA to be analysed is amplified by long-range PCR. Fluorescently labelled RNA is then prepared by *in vitro* transcription and this is hybridized to the array. The hybridization patterns are imaged with a high-resolution confocal scanner and a typical result is shown in Fig. 5.17. For the purpose of resequencing the human mitochondrial genome (L = 16 569 bp) with a tiled array of $P^{15,7}$ probes then a total of 66 276 probes ($4 \times 16\,569$) of the possible $\sim 10^9$ 15mers would be required.

Despite the fact that hybridization has not, as yet, been used for *de novo* sequencing the technique is still of great value. First, it can be used very easily to analyse genomes for polymorphisms (Goffeau 1997). For example, Chee *et al.* (1996) prepared a $P^{25,13}$ tiling array for the mitochondrial genome. This consisted of 136 528 synthesis cells, each 35 μm square in size. In addition to a 4L tiling across the genome, the array contained a set of probes representing a single-base deletion at every position across the genome and sets of probes designed to match a range of specific mtDNA haplotypes. After hybridization of fluorescently labelled target RNA 99% of the sequence could be read correctly simply by identifying the highest intensity in each column of four substitution probes. The array also was used to detect three disease-causing mutations in a patient with hereditary optic neuropathy (Fig. 5.18). A refinement to this method also has been developed. Ideally, the hybridization signals from the reference and test

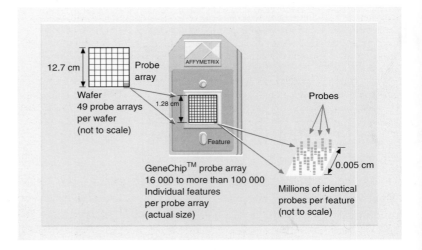

Figure 5.16 Structure of a sequencing microchip. (Courtesy of Dr M. Chee, Affymetrix.)

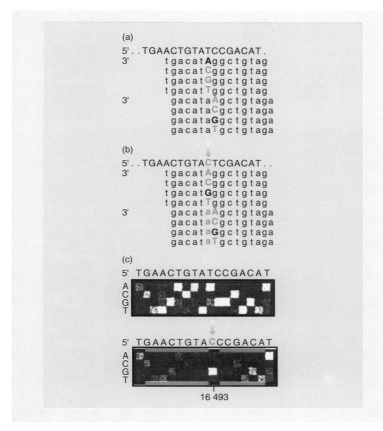

Figure 5.17 Design and use of a 4L tiled array. Each position in the target sequence (uppercase letters) is queried by a set of four probes on the chip (lowercase letters), identical except at a single position, termed the substitution position, which is either A, C, G or T (bold black indicates complementarity, green a mismatch). Two sets of probes are shown, querying adjacent positions in target. (b) Effect of a change in the target sequence. The probes are the same as in (a), but the target now contains a single-base substitution (base C, shown in green and arrowed). The probe set querying the changed base still has a perfect match (the G probe). However, probes in adjacent sets that overlap the altered target position now have either one or two mismatches (green) instead of zero or one, because they were designed to match the target shown in (a). (c) Hybridization to a 4L tiled array and detection of a base change in the target. The array shown was designed to the mt1 sequence. (Top) Hybridization to mt1. The substitution used in each row of probes is indicated to the left of the image. The target sequence can be read 5′ to 3′ from left to right as the complement of the substitution base with the brightest signal. With hybridization to mt2 (bottom), which differs from mt1 in this region by a T → C transition, the G probe at position 16 493 is now a perfect match, with the other three probes having single-base mismatches (**A** 5, **C** 3, **G** 37, **T** 4 counts). However, at flanking positions, the probes have either single- or double-base mismatches, because the mt2 transition now occurs away from the query position. (Reproduced with permission from Chee *et al.* 1996, © American Association for the Advancement of Science.)

DNAs should be compared by hybridization to the same array. This can be done by using a two-colour labelling and detection scheme in which the reference is labelled with phycoerythrin (red) and the target with fluorescein (green). By processing the reference

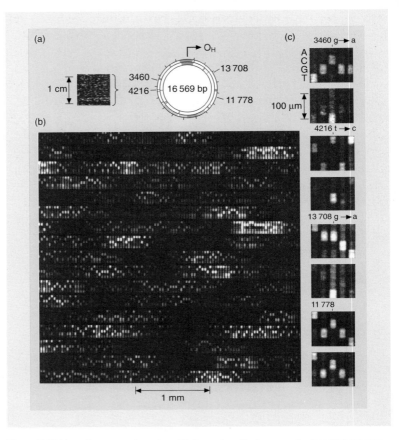

Figure 5.18 Use of a sequencing microchip to analyse the human mitochondrial genome. (a) An image of the array hybridized to 16.6 kb of mitochondrial target RNA (L strand). The 16 569 bp map of the genome is shown, and the H strand origin of replication (O_H), located in the control region, is indicated. (b) A portion of the hybridization pattern magnified. In each column there are five probes. A, C, G, T and Δ, from top to bottom. The Δ probe has a single base deletion instead of a substitution and hence is 24 instead of 25 bases in length. The scale is indicated by the bar beneath the image. Although there is considerable sequence-dependent intensity variation, most of the array can be read directly. The image was collected at a resolution of $\sim$ 100 pixels per probe cell. (c) The ability of the array to detect and read single base differences in a 16.6 kb sample is illustrated. Two different target sequences were hybridized in parallel to different chips. The hybridization patterns are compared for four positions in the sequence. Only the $P^{25.13}$ probes are shown. The top pannel of each pair shows the hybridization of the mt3 target, which matches the chip P^0 sequence at these positions. The lower panel shows the pattern generated by a sample from a patient with Leber's hereditary optic neuropathy (LHON). Three known pathogenic mutations, LHON3460, LHON4216, and LHON13 708, are clearly detected. For comparison, the fourth panel in the set shows a region around position 11 778 that is identical in both samples. (Reproduced with permission from Chee *et al.* 1996, © American Association for the Advancement of Science.)

and target together, experimental variability during the fragmentation, hybridization, washing and detection steps is eliminated.

The most exciting feature of the above is that it is carried out extremely fast and with relatively little labour requirement; for example, five chips can be read per hour which equates to 40–50 analyses per day. By comparison, a conventional gel-based sequencer with 48 lanes run twice a day might permit two mitochondrial sequences to be analysed in the same period. Furthermore, there are significant reductions in sample preparation time because the entire genome is labelled in a single reaction equating to a single gel-based sequencing reaction. Also, sequence analysis at the level of data analysis is automated, sequences being read in a matter of minutes. Note that with conventional dideoxy sequencing an accuracy of 99.9% will result in errors in over 66% of genes (see p. 131). One benefit of the rapidity of microchip sequencing is that it could be used to confirm quickly and cheaply the fidelity of sequence data derived by gel-based methods.

6 Analysing sequence data

DNA sequence databases

Since the current DNA sequencing technology was developed a large amount of DNA sequence data has accumulated. These data are maintained in three databases: the National Center for Biotechnology Information in the USA, the DNA Databank of Japan and the European Bioinformatic Institute in the UK (Benson *et al*. 1996, 1997; Stoesser *et al* 1997; Tateno & Gojobori 1997). Each of these three groups collects a portion of the total sequence data reported worldwide and all new and updated database entries are exchanged between the groups on a daily basis. In addition, several specialized genome databases exist, including seven for bacterial genomes: four for *E. coli*, two for *Bacillus subtilis* and one at the Institute for Genome Research, an organization responsible for the complete sequencing of a number of genomes. Users worldwide can access these databases directly via the Worldwide Web or receive the information on CD-ROMs. The former option is the best because it ensures that an up-to-date database is being used. There are a number of different sequence retrieval systems and the best of these are Network Entrez and DNA Workbench (Brenner 1995).

As of late 1996, there were over 900 000 nucleotide records available from the databanks, an increase of 100% over the same time a year before. However, most of this increase came from the submission of large numbers of partial cDNA sequences (ESTs) and ESTs now account for more than half of all records (Leipe 1996). Most of these ESTs are from humans and mice. Of the ~ 20 000 organisms represented in the databases with one or more sequences, about 69% are eukaryotes (Fig. 6.1). By far the largest number of non-EST sequence records are available for mammals and 80% of these are derived from human DNA. A strongly biased distribution of molecular information also is apparent if one considers the average number of sequence records that are available for every organism that is represented in the databases. There are, on average, one to two orders of magnitude more records for mammals, mostly humans and mice, than for any other group of organisms (Fig. 6.2). As well as nucleotide sequence data these databases also store information on protein sequences.

In working with databases it is important to recognize their inherent deficiencies. As well as errors in the sequencing process itself, there can be transcriptional errors when the data is transferred from laboratory notebook to publications and databases. For example, when screening 300 human protein sequences in the SWISS-PROT database that had been published separately more than once, Bork (1996) found that 0.3% of the amino acids were different. This is a lower limit, for many corrections will already have been made and in many instances the sequences appearing in two different publications are not independent. Note that only stop codons and frameshifts can be detected unambiguously: point mutations are hard to verify as natural polymorphisms or strain differences cannot easily be excluded. Sequencing by hybridization may be of great use here. Other database problems include misspelling of genes resulting in confusion with ones of similar name or the generation of synonyms: different genes being given the same name and multiple synonyms for the same gene. Examples of the latter are the *E. coli* gene *hns* which has eight synonyms and the protein annexin V which has five synonyms.

Analysing sequence data

Discovering new genes and their functions is a key step in analysing new sequence data. The process is facilitated by special-purpose gene-finding software, by searches in key databases and by programs for finding particular sites relevant to gene expression, e.g. splice sites and promoters. Unfortunately no one software

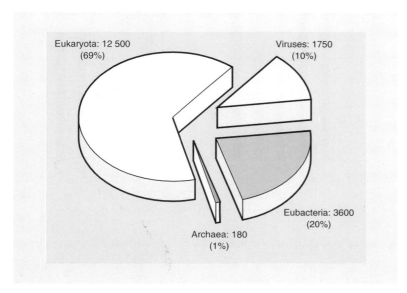

Fig. 6.1 Number of organisms represented with one or more sequence in the DNA databases. (Redrawn with permission from Leipe 1996, © *Current Opinion in Genetics and Development.*)

package contains all the necessary tools. Rather, optimal gene finding is dependent on combining evidence derived from use of multiple software tools (Table 6.1).

Fickett (1996) has described a framework for finding genes which makes use of a number of different software programs. Evidence for the presence or absence of genes in a sequence is gathered from a number of sources. These include sequence similarity to other features such as repeats which are unlikely to overlap protein coding sequences, sequence similarity to other genes, statistical regularity evincing apparent codon bias over a region, and matches to template patterns for functional sites on the DNA such as consensus sequences for TATA boxes. All the information so gathered is integrated to make as coherent a picture as possible.

When analysing sequences from eukaryotes it is best to locate and remove interspersed repeats before searching for genes. Not only can repeats confuse other analyses such as database searches but they provide important negative information on the location of gene features. For example, such repeats rarely overlap the promoters or coding portions of exons. Once this is done the next step is to identify open-reading frames (ORFs). Despite the availability of sophisticated software search routines, unambiguous assignment of ORFs is not easy. For example, the *Haemophilus influenzae* genome sequence submitted to the database included 1747 predicted protein-encoding genes (Fleischmann *et al.* 1995).

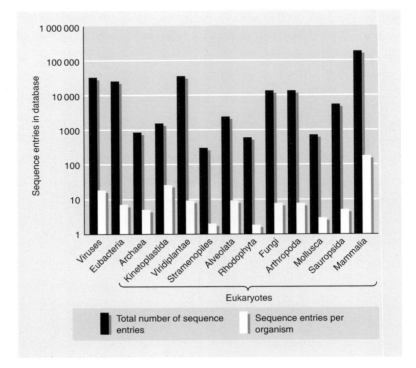

Fig. 6.2 Distribution of non-EST nucleotide sequence records among major groups of organisms. The left column of each pair represents the total number of records and the right column represents the average number of records for every organism that is represented in the database. Note that the scale of the Y-axis is logarithmic. (Redrawn with permission from Leipe 1996, © *Current Opinion in Genetics and Development.*)

When Tatusov *et al.* (1996) re-analysed the sequence data using different algorithms and different discrimination criteria, they identified a new set of 1703 putative protein-encoding genes. In addition to 1572 ORFs, which remained the same, they identified 23 new ORFs, modified 107 others and discarded the balance. Note that gene finding is relatively easy in compact and almost intron-free genomes such as yeast. In higher plants and animals the task is much greater, for a 2 kb ORF could be split into 15 exons spread over 30 kb of genomic DNA.

Database searches

Searching for a known homologue is the most widely used means of identifying genes in a new sequence. If a putative protein encoded by an uncharacterized ORF shows statistically significant similarity to another protein of known function, this simultaneously proves, beyond doubt, that the ORF in question is a *bona fide* new gene and predicts its likely function. Even if the homologue of the new protein has not been characterized, useful information is produced in the form of conserved motifs that may

Table 6.1 Internet tools for gene discovery in DNA sequence data. (Reproduced with permission from Fickett 1996)

Category	Service	Organism(s)	Address
Repeat analysis	Pythia; give a list of repeats in sequence	Human	pythia@anl.gov
	Repbase; repeat collections	Human and several other collections	ftp;//ncbi.nlm.nih.gov; repository/rebase/REF
	BLASTX; tools to mask repeat occurrences	Any	ftp://ncbi.nlm.nih.gov; pub/jmc
Database search	BLAST; search sequence databases	Any	blast@ncbi.nlm.nih.gov
	FASTA; search sequence databases	Any	fasta@ebi.ac.uk
	BLOCKS; search for functional motifs	Any	blocks@howard.fhcrc.org
	ProfileScan	Any	http://ulrec3.unil.ch/software/ PFSCAN_form.html
	MotifFinder	Any	motif@genome.ad.jp
Gene identification	FGENEH; integrated gene identification	Human	service@theory.bchs.uh.edu
	GeneID; integrated gene identification	Vertebrate	geneid@bir.cedb.uwf.edu
	GeneMark; coding region identification	Many individual species	genemark@ford.gatech.edu
	GeneParser; integrated gene identification	Human	http://beagle.colorado.edu/ ~eesnyder/GeneParser.html
	GenLang; integrated gene identification	Dicots, *Drosophila*, vertebrates	genlang@cbil.humgen.upenn.edu
	GRAIL integrated gene identification	Human	grail@ornl.gov (also graphical interface)
	EcoParse; integrated gene identification	*Escherichia coli*	ecoparse@cse.ucsc.edu
'Signal' recognition	PromoterScan	Eukaryotes	Contact Dan Prestridge at danp@biosci.cbs.umn.edu for FTP
	NetGene	Human	netgene@virus.fki.dth.dk

be important for protein function. In this way, Koonin *et al*. (1994) analysed the information contained in the complete sequence of yeast chromosome III and found that 61% of the probable gene products had significant similarities to sequences in the current databases. As many as 54% of them had known functions or were related to functionally characterized proteins and 19% were similar to proteins of known three-dimensional structure. Other examples are given later. The methods of choice for screening databases in this way are those such as BLASTX that translate the query nucleotide sequence in all six reading frames and compare the resulting putative protein sequences to the protein sequence database. Such methods allow the detection of frameshift errors and will not miss even small ORFs if homologues are present in the database.

Significant matches of a novel ORF to another sequence may be in any of four classes (Oliver 1996a). First, a match may predict both the biochemical and physiological function of the novel gene. An example is ORF YCR24c from yeast chromosome III and the *E. coli* Asn-tRNA synthetase. Second, a match may define the biochemical function of a gene product without revealing its biochemical function. An example of this are five protein kinase genes found on yeast chromosome III whose biochemical function is clear (they phosphorylate proteins) but whose particular physiological function in yeast is unknown. Third, a match may be to a gene from another organism whose function is unknown in that organism, e.g. ORF YCR63w from yeast, protein G10 from *Xenopus* and novel genes from *Caenorhabditis* and humans, none of whose function is known. Finally, a match may occur to a gene of known function that merely reveals that our understanding of that function is very superficial, e.g. yeast ORF YCL17c and the Nif S protein of nitrogen-fixing bacteria. After similar sequences were found in a number of bacteria which do not fix nitrogen it was shown that the Nif S protein is a pyridoxal phosphate-dependent aminotransferase.

Because the DNA and protein sequence databases are updated daily it could be that homologues of previously unidentified proteins have been found. Thus it could pay to repeat the search of the databases at regular intervals as Robinson *et al*. (1994) found. Starting with more than 18 million bp of prokaryotic DNA sequence they executed a systematic search for previously undetected protein-coding genes. They removed all DNA regions known to encode proteins or structural RNAs and used an algorithm to translate the remaining DNA in all six reading frames. A search of the resultant translations against the protein sequence databases uncovered more than 450 genes which previously had escaped detection. Seven of these genes belonged to gene families not previously identified in prokaryotes. Others belonged to gene families with critical roles in metabolism. Clearly, periodic exhaustive reappraisal of databases can be very productive!

If homology with known proteins is to be used to analyse sequence data, then sequence accuracy is paramount. With the gene density and ORF size distribution of yeast (Dujon *et al.* 1994) and the nematode (Wilson *et al.* 1994), even relatively rare sequencing errors, many of which are missing nucleotides, will result in a large fraction of the protein-coding genes being affected by frameshifts. At 99% accuracy virtually all genes will contain errors and at 99.9% accuracy two-thirds of the genes will still contain an error, most probably a frameshift. With the 99.97% accuracy obtained in the yeast sequencing about one-third of predicted genes will still contain errors (Fig. 6.3).

Because sequence similarity search programs are so vital to the analysis of DNA sequence data, much attention has been paid to the precise algorithms used and their precise speed. However, the effectiveness of these searches is dependent on a number of factors.

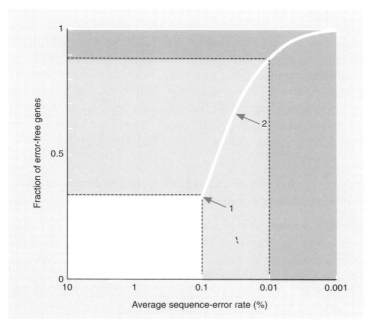

Fig. 6.3 Quantitative effect of sequence accuracy on gene accuracy. The figure indicates the fraction of error-free genes predicted from a DNA sequence, as a function of average sequence accuracy. The theoretical curve was computed using the size distribution of protein-coding genes in yeast and assuming random occurrence of sequencing errors. All types of errors are considered together. In practice, frameshifts, which have the most deleterious effects on gene interpretation, represent about two-thirds of all sequencing errors. The figure illustrates the difference between an average sequence accuracy of 99.9%, where only one-third of all genes are properly described, and an average sequence accuracy of 99.99%, where more than 85% of genes are properly described. It also illustrates the difference between average database entries (99.9% accuracy, arrow 1) and systematic sequencing programs with sequence verifications (99.97% accuracy, arrow 2). (Redrawn with permission from Dujon 1996.)

This include the choice of scoring system, the statistical significance of alignments and the nature and extent of sequence redundancy in the databases (Altschul *et al.* 1994; Coulson 1994). Most database search algorithms rank alignments by a score whose calculation is dependent upon a particular scoring system. Optimal strategies for detecting similarities between DNA protein-coding regions differ from those for non-coding regions. Special scoring systems have been developed for detecting frameshift errors in the databases, a problem highlighted above. Thus a database search program should make use of a variety of scoring systems. Furthermore, given a query sequence, most database search programs will produce an ordered list of imperfectly matching database similarities, but none of them need have any biological significance. Many of the genes originally predicted by these statistical methods subsequently have proved to be homologous to newly described genes or have been confirmed experimentally, thus supporting the robustness of the predictive methods (Koonin *et al.* 1996).

Eventually, with the accumulation of new sequences, sequence conservation will become the definitive criterion for gene identification and the contribution of statistical methods will decrease. Nevertheless, it is still likely that some genes will not have identifiable homologues and statistical and experimental approaches will remain necessary for their detection. Furthermore, even for genes that have homologues, statistical methods of coding-potential analysis will remain useful for localizing frameshifts and choosing among the possible initiation codons.

Sequence analysis at the whole genome level: BCRs, ACRs, orthologues and paralogues

Now that the complete genome sequence is available for a number of different organisms, including prokaryotes with genome sizes ranging from 580 kb (*Mycoplasma genitalium*) to 4.7 Mb (*E. coli*) and one eukaryote (yeast), it is possible to undertake comparisons at the whole organism level. This takes two forms. At one level it is possible to ask questions about evolutionary relationships; at the other one can attempt to predict key biological activities, such as metabolism, of a poorly characterized organism by comparison with a well-understood one.

In Eubacteria at least, most of the bacterial proteins appear to be highly conserved in evolution. This sequence conservation is not due to trivial similarity to homologues from closely related bacterial species. For example, about 65% of the *E. coli* proteins contain bacterial conserved regions (BCRs). That is, regions conserved at least at the level of distantly related bacteria. Over 40% contain regions shared with eukaryotic or archeal homologues, i.e. ancient conserved regions (ACRs). Analysis of the *H. influenzae* and *M. genitalium* predicted protein sequences shows that the proportions of proteins containing ACRs and BCRs are very similar despite the very different coding potentials (Koonin *et al.* 1996). Thus, smaller genomes are not enriched for conserved sequences. This observation implies that a 'minimal' genome, such as that of *M. genitalium* (580 kb) is not composed solely of highly conserved 'housekeeping' genes. Rather, it suggests that bacteria maintain a balance between highly conserved and more variable genes, even while dramatic changes in genome size are taking place.

In order to compare genome organization in different organisms it is necessary to distinguish between *orthologues* and *paralogues*. Orthologues are homologous genes in different organisms that encode proteins with the same function and that have evolved by direct, vertical descent. Paralogues are homologous genes within an organism encoding proteins with related but non-identical functions. About 50% of *E. coli* proteins belong to clusters of paralogues. Most of these groups are small but there are several large clusters consisting of proteins involved in metabolite transport and gene regulation. The *H. influenzae* genome is 62% smaller than that of *E. coli* and the *M. genitalium* genome is 88% smaller and there is a corresponding reduction in the fraction of paralogue clusters: 35% for *H. influenzae* and 25% for *M. genitalium* (Koonin *et al.* 1996). That is, the smaller the genome size, the lower the level of gene paralogy. In some cases this reduction in paralogy is due to the absence of entire functional systems. For example, the *M. pneumoniae* genome (816 kb) encodes a restriction-modification system, two transport systems for carbohydrates and three genes encoding the arginine dihydrolase pathway which are absent from the smaller (580 kb) genome of *M. genitalium* (Himmelreich *et al.* 1997). In others it is due to clusters containing a smaller number of proteins (Fig. 6.4). In some cases there are

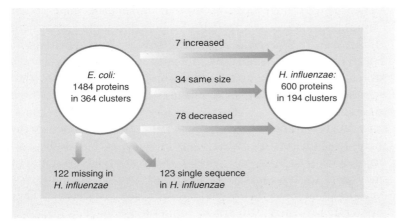

Fig. 6.4. Comparison of paralogous protein clusters in *Escherichia coli* and *Haemophilus influenzae*. (Redrawn with permission from Tatusov *et al.* 1996.)

multiple protein species in *E. coli* but only one in the other organisms.

Tatusov *et al*. (1996) noted significant sequence similarity to *E. coli* proteins for 1307 putative *H. influenzae* proteins and, of these, 1128 appeared to have an *E. coli* orthologue. The percentage identity between orthologues spanned a wide range, between 18% and 98%, but the majority of pairs contained between 40% and 80% identical amino acid residues. On average, the *H. influenzae/ E. coli* orthologues contained 59% identical and 75% similar amino acids.

The lists of orthologues derived for *H. influenzae* and *M. genitalium* has provided a basis for understanding the differences between the organisms at the gross cellular/biochemical level. For example, the functional categories most highly conserved are the components of the replication, transcription and translation machinery and molecular chaperones. The representation of all other functional categories of proteins is dramatically reduced in *M. genitalium*. By contrast, in *H. influenzae* proteins involved in such housekeeping functions as DNA repair and nucleotide biosynthesis are largely preserved. In *M. genitalium* there is a virtual absence of proteins which, in other bacteria, are involved in the regulation of gene expression. These have been preserved in *H. influenzae*. Perhaps most interesting of all, analysis of orthologues has enabled Koonin *et al*. (1996) and Tatusov *et al*. (1996) tentatively to elucidate the principal metabolic pathways of *H. influenzae* and *M. genitalium* in the absence of any experimentally-derived biochemical data. A similar analysis on the archeon *Methanococcus jannaschii* (Bult *et al*. 1996) has shown that the majority of its genes related to energy conversion, cell division and metabolism are most similar to those found in bacteria. By contrast most of the genes involved in transcription, translation and replication are more similar to those of eukaryotes.

In some instances the identification of orthologues reveals our gross lack of knowledge. A good example is provided by the *Nramp* gene (Cellier *et al*. 1996). In humans this gene confers natural resistance or susceptibility to infection with a wide range of microbes. In *Drosophila*, if it is defective, it causes a behavioural defect in taste discrimination. In yeast it may be involved in protein import and vacuolar metabolism of calcium. Finally, three homologues with different expression patterns have been found in the roots of rice.

By analysis of orthologues it is possible to determine whether there are core sets of genes. Clayton *et al*. (1997) looked for orthologue sets that were present in all complete genome sequences. They found 53 groups that were present in a eukaryote (yeast), an archean (*M. jannaschii*) and five eubacteria. These represent a core set of genes common to all the organisms surveyed but do not represent the full core set. Eighty orthologue groups consisted of genes from yeast and *M. jannaschii* which are absent from the eubacteria analysed. Many of these are involved in information processing: replication, transcription and translation. Similarly, there are over 500 orthologue sets common to all the eubacteria analysed which are absent from yeast and *M. jannaschii*.

Orphan genes

Orphan genes are of two types. Single orphans, also known as sequence orphans, have no sequence relatives in the databases. Orphan pairs have sequence homologues but their function is unknown. Orphan genes have been identified in all the major genome sequencing projects. For example, Oliver *et al.* (1992) found that approximately half of the protein-coding ORFs identified from the complete sequence of yeast chromosome III had no orthologues or paralogues. Analysis of the complete yeast genome sequence identified 20–30% of ORFs as being orphan (Casari *et al.* 1996; Dujon 1996) and about half of these were orphan pairs. In *Mycoplasma pneumoniae*, Himmelreich *et al.* (1996) found that, of the ORFs identified, 9.9% were single orphans and 6.7% were orphan pairs. Since orphans are defined by the absence of known function and of structural homologues of known function, it seems likely that they will vanish in time as more data accumulates from the many different sequencing projects now underway.

Even in the absence of homologues, computers can provide some clues about the nature of some orphan genes by identifying amino acid sequences common to particular protein structural motifs, e.g. zinc fingers, ATP-binding sites, etc. As an example, approximately one-third of the orphan genes from yeast chromosome III have sequences characteristic of transmembrane helices. A significant proportion are predicted to specify more than four transmembrane helices, suggesting that they might encode transporters.

From computer analysis to gene function

Computer comparisons of genome sequences produce conclusions that are important in their own right: for example, genome organization conservation and possible evolutionary events. Ultimately, however, the functions of all the protein-coding ORFs need to be identified. The principal way of achieving this is to inactivate the gene product by gene disruption, preferably by deletion of the entire ORF. Information about the biological function can then be inferred by monitoring the fitness of the deletion strain under a variety of selective growth conditions. Unfortunately, the time and labour involved in analysing

CHAPTER 6
Analysing sequence data

individual deletion strains makes it difficult to apply this approach on a genome-wide level. Strategies have been reported in which gene disruptions and selections are performed *en masse* using randomly integrated transposons (Smith *et al*. 1995). However, thousands of mutants need to be individually examined to identify those with modified growth characteristics; this again rules out a genome-wide search. Recently, Shoemaker *et al*. (1996) have described a quantitative and highly parallel method for analysing large numbers of deletion mutants of *Saccharomyces cerevisiae*.

Two novel approaches, termed 'quantitative phenotypic analysis' and 'molecular bar-coding', are used in the method of Shoemaker *et al*. (1996). In this, directed gene replacement is used to generate individual deletion strains which are each labelled with a distinguishing 20 bp molecular tag. Tagged strains are pooled and analysed in parallel through selective growth conditions. The level at which each strain survives a given competitive growth condition can be determined by hybridizing the tags to high-density oligonucleotide arrays on microchips (see p. 122) This approach affords four benefits. First, large pools of tagged deletion strains can be simultaneously analysed because of the sensitivity provided by the microchip arrays. Second, deletion strains have ORFs completely deleted, avoiding residual or altered functions associated with truncated products. Third, deletion strains with interesting phenotypes are directly identified without need for further cloning or sequencing. Finally, the quantitative nature of the chip hybridization assay should make it possible to reveal subtle phenotypes that otherwise may be missed.

The molecular tags used are 20mers specifically designed to serve as unique identifiers. Tag sequences need to be as different as possible yet still retain similar hybridization properties to facilitate simultaneous analysis on the microchips. Using an algorithm, Shoemaker *et al*. (1996) were able to select a set of 9105 maximally distinguished tags that are predicted to have similar melting temperatures, no secondary structure and at least five mismatches with all other tags in the set. This set of tags is sufficient to analyse all putative protein-coding genes (~ 6000) in yeast.

For gene analysis in yeast, one of these tags is linked to a kanamycin-resistance gene and yeast sequences on either side of the ORF to be deleted (Fig. 6.5). This cassette then is transformed into a haploid yeast strain and homologous recombination results in deletion of the targeted ORF and its replacement with the 20mer tag. Large numbers of individually tagged deletion strains are pooled and grown competitively under various selective conditions. The molecular tags are amplified from the surviving strains using a common set of primers and hybridized to a high-density array containing, at defined positions, known oligonucleotides, which are complementary to the tag sequences.

Finally, the intensities of the hybridization signals for the tags on the array are measured and used to determine the relative abundance of the corresponding deletion strains in the pool. The feasibility of this method has been proven and the only real disadvantage is the initial labour-intensive step of making the deletion mutants.

There are two generic problems associated with identifying gene function by analysis of deletion mutants (Johnston 1996). First, most mutants do not display obvious phenotypes. The fraction of mutants manifesting a phenotype undoubtedly will increase if systematically analysed and Oliver (1996b) has described such a program. Nevertheless, a significant proportion of genes will resist such examination. It could be that some of these genes are expressed as proteins but are in fact dispensable and have not yet decayed from the genome. The human α- and β-globin genes may be examples of such 'expressed pseudogenes'. Alternatively, the right selective environment has not been found. Second, and more problematic, the phenotypes that many mutants exhibit, e.g. failure to grow in particular media or under particular growth regimens, will not necessarily give information about function.

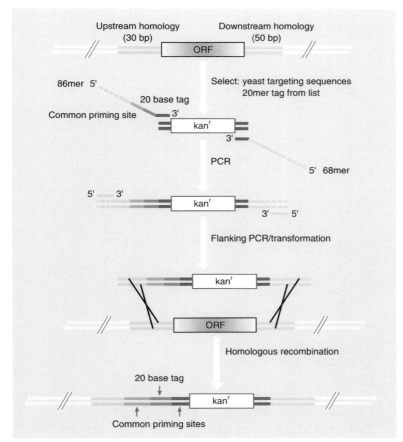

Fig. 6.5. Targeting strategy used to generate tagged deletion strains of yeast. A dominant selectable marker (kan[r]) is amplified using a pair of long primers that contain yeast sequences on the 5′ end and homology to the marker on the 3′ end. One of the oligonucleotides is a 68mer which contains 50 bases of yeast homology and 18 bases of homology to the marker. The other oligonucleotide is an 86mer which contains a 20 base tag and an 18 base tag priming site in addition to the 30 bases of yeast homology and the 18 bases of marker homology. A second round of PCR is performed with 20mers which are homologous to the ends of the initial PCR product to increase the amount of full length product. The product is transformed into a haploid yeast strain and homologous recombination results in the replacement of the targeted ORF with the marker, 20mer tag and tag priming site. (Redrawn with permission from Shoemaker *et al.* 1996, courtesy of *Nature Genetics*.)

Nevertheless, a significant amount of useful information will be generated in a cost-effective way.

Specialized databases

Many different specialized databases exist, e.g. databases for radiation hybrids, small RNAs, androgen receptor genes, immunogenetics, viroids and P450 genes to name but a few. The 1 January 1997 issue of *Nucleic Acids Research* (vol. **25**, no. 1) has summaries of 64 of these specialized databases.

7 Finding genes in large genomes

Introduction

Using classic genetics it is relatively easy to show the mode of inheritance of a particular Mendelian trait. Developing an understanding of the biological basis of the phenotype is harder, for this requires a detailed knowledge of the appropriate gene(s), genetic control of the gene(s), identification of gene products and their role in the life of the host. In most instances this requires gene isolation using recombinant DNA technology. Where biochemical information is available, it usually is possible to devise a suitable protocol for selecting the appropriate gene from gene libraries (see Old & Primrose 1994). This has been described as *functional cloning* (Collins 1992). However, for the vast majority of traits, including over 4000 in man, no biochemical information is available. Recently a number of techniques have been developed for *positional cloning* of genes of unknown function. Functional cloning depends upon application of biological information, while positional cloning is initiated by mapping the responsible gene to its correct location on a chromosome (Fig. 7.1). Successive narrowing of the candidate interval eventually results in the identification of the correct gene whose function then can be studied.

Mapping the gene of interest

In practice, one starts with a collection of pedigrees in which the responsible gene is segregating. These families are studied with multiple polymorphic markers until evidence for chromosomal location is obtained. Then linkage to other markers on that chromosome is determined. Clearly, the greater the number of markers available per chromosome, the more useful will be the mapping data. The closer the gene of interest to a known marker, the easier will be the later stages. In yeast, with little repeated DNA, the separation of the two will not be great for 1 cM is equivalent to 3 kb of DNA sequence. In higher eukaryotes the separation will be very much greater; for example, in man 1 cM translates to 1 Mb of DNA sequence and this underscores the desire to have a detailed map with markers spaced every 100 kb (see p. 3). Absolute location of the gene of interest is facilitated if the corresponding trait is associated with chromosome abnormalities that are cytogenetically visible, e.g. fragile sites, deletions, duplications and translocations.

The next stage is chromosome walking (p. 64) and jumping to bridge the gap between the nearest markers and the gene of interest. Although attractive in theory, chromosome walking suffers from the problems that it is very laborious and long-range walking is not practicable. In addition, the presence of unclonable DNA can bring the walk to a grinding halt. Chromosome jumping avoids these pitfalls. This involves circularization of very large genomic DNA fragments followed by cloning DNA from the region covering the closure site of these circles. This brings together DNA sequences that were originally located a considerable distance apart in the genome. These cloned DNAs from the closure sites make up a 'jumping library'. Figure 7.2 shows a strategy (Poustka *et al.* 1987) for creating a jumping library using *Not*I-digested human genomic DNA. This enzyme cuts only rarely in mammalian DNA and so drastically reduces the size of the library needed to cover the mammalian genome. This makes analysis of jumps easy if PFGE and Southern blotting of *Not*I-digested genomic DNA are used. However, with a complete digest such as this, overlaps between neighbouring *Not*I fragments have to be obtained indirectly. One solution is to use 'linking clones'

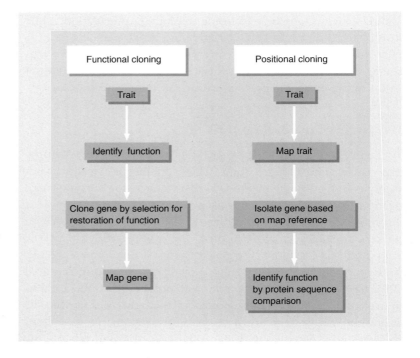

Figure 7.1 Comparison of functional and positional cloning. Functional cloning depends on the availability of information about the protein product and/or function of the responsible gene. Mapping then follows cloning. Positional cloning moves in the opposite direction. Here mapping is essential to the process and function is determined after the gene is identified.

containing conventionally cloned fragments with an internal *Not*I
site. A variant of the strategy shown in Fig. 7.2 employs partial
digestion with a restriction endonuclease to generate large genomic DNA fragments. This technique has been applied to a jump
of 100 kb in the cystic fibrosis locus (Collins *et al.* 1987) and to a
jump of 200 kb in the region of the Huntington's disease gene
(Richards *et al.* 1988).

Identifying the gene of interest

Once the locus of the gene of interest has been mapped, one or
more YAC clones are isolated which contain sequences from the
surrounding region and these are sub-cloned to generate a contig
of overlapping phage or cosmid clones from which DNA can be
easily purified. Several different approaches are available for gene

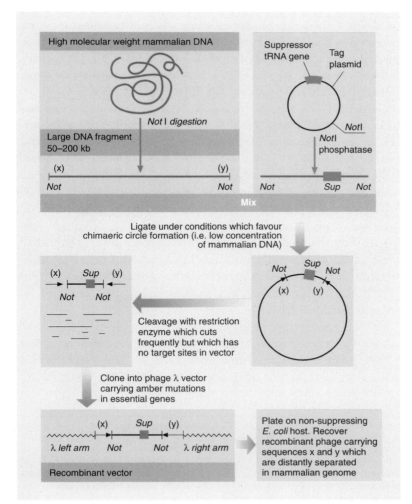

Figure 7.2 Construction of a
jumping library using *Not*I-
digested mammalian DNA. The
suppressor tRNA gene acts as
a selective marker for the
recombinants because it
suppresses the amber
mutations in essential genes
of the phage vector.

141

identification (Table 7.1) and these are reviewed by Monaco (1994).

One approach is based on the observation that coding sequences are strongly conserved during evolution, whereas non-coding DNA is not (see, for example, the *Drosophila* mapping experiment described on p. 68). A DNA clone that may contain a gene can be hybridized against a Southern blot of genomic DNA samples from a variety of species, a so-called zoo blot (Monaco *et al.* 1986). At reduced hybridization stringency probes containing human genes, for example, will generally show strong hybridization to animal genomic DNA. Probes containing only human non-coding sequences will not. However, genes that are species-specific will not be detected in this way. By means of such zoo blots Sedlacek *et al.* (1993) isolated genomic DNA species conserved between humans mice and pigs.

One important approach in the search for genes is the isolation of CpG island DNA (see p. 41). The DNA adjacent to rare-cutter sites, which are concentrated in CpG islands (e.g. *Not*I, GCGGCCGC), has been cloned from a number of regions of the human genome. Analysis of such clones has shown that they are derived from CpG islands. The clones detect expressed sequences and have been used to identify a number of novel genes in large, cloned genomic regions (Cross & Bird 1995).

Conserved DNA fragments, CpG island-containing fragments and random single-copy DNA fragments can be tested for expression by hybridization to Northern blots of RNA isolated from fetal and adult tissues. If a transcript is identified then the genomic DNA fragment subsequently can be hybridized to cDNA libraries. Although Northern blotting and the analysis of conserved DNA fragments and CpG islands have been particularly effective in the isolation of many human disease genes (Table 7.2) they are very labour intensive. An alternative approach is to directly hybridize whole YAC or cosmid inserts to cDNA libraries on filters after suppression of repeated sequences in the probe. Although technically difficult, the method has enabled many new genes to be isolated (Monaco 1994).

Another approach to gene isolation is cDNA selection (Lovett *et al.* 1991; Parimoo *et al.* 1991). In this method, an amplified cDNA library is hybridized to immobilized YAC or cosmid clones that cover part of the genome of interest. Of those cDNAs

Table 7.1 Strategies to identify genes during positional cloning

Traditional approaches	Newer approaches
Conservation on zoo blots	Exon amplification
CpG islands	cDNA selection
Northern blots	Genomic sequencing and computer analysis
cDNA library	Regionally mapped candidate genes

Table 7.2 Some human disease genes identified by positional cloning

Disease locus	Cytogenetic rearrangement
Chronic granulomatous disease	+
Duchenne muscular dystrophy	+
Retinoblastoma	+
Wilms tumour	+
Cystic fibrosis	−
Neurofibromatosis type 1	+
Testis determining factor	+
Fragile X syndrome	+
Familial polyposis coli	+
Choroideremia	+
Kallmann syndrome	+
Aniridia	+
Myotonic dystrophy	−
X-linked agammaglobulinaemia	−

isolated, at least one should correspond to the desired gene. Again, the process is technically difficult (Lovett 1994) because of simultaneous selection of cDNAs with homology to pseudogenes and low copy number repeated sequences. A variation of this method is to use exon amplification (Buckler *et al*. 1991). In this method, random segments of chromosomal DNA are inserted into an intron present within a mammalian expression vector. After transfection, cytoplasmic mRNA is screened by PCR amplification for the acquisition of an exon from the genomic fragment. The amplified exon is derived from the pairing of unrelated vector and genomic splicing signals. A number of human genes have been isolated successfully using this method (Monaco 1994).

As noted in Chapter 5, a number of groups have begun large-scale sequencing of large genomes and analysis of assembled sequences for potential ORFs. If the gene of interest lies within a sequenced region then its identification should be relatively easy. The approach of sequencing genomic DNA to find expressed sequences has been applied to the positional cloning of the human gene for Kallmann syndrome. Whereas one group used conservation of sequences from YACs to identify potential exons (Franco *et al*. 1991), the other completely sequenced 60 kb of genomic DNA and identified potential coding regions (Legouis *et al*. 1991).

Isolation and characterization of the muscular dystrophy (DMD) gene as an example of positional cloning

The first successful application of positional cloning was the elucidation of the molecular basis of Duchenne muscular dystrophy (Hoffman *et al*. 1987). In this instance, isolation of the gene was facilitated by the availability of affected individuals with chromosomal translocations and deletions which localized the gene within the p21 region of the human X chromosome. The availability of a patient with a deletion was particularly fortuitous as it enabled subtraction cloning to be undertaken. For this, genomic DNA from a normal individual was digested with endonuclease *Mbo*I which produces fragments with GATC overhangs. The *Mbo*I fragments were denatured and annealed with a 200-fold excess of denatured DNA from the deletion patient and which had been sonicated to produce ragged ends. Under these circumstances Xp21 DNA present in normal individuals but absent in the deletion patient will anneal with itself and will be the only DNA with a GATC overhang at each end. Thus it could be selectively cloned in a vector cut with *Bam*HI. Individual DNA clones then were used as probes in Southern blot hybridizations against normal and patient DNA samples to identify those which could be carrying the DMD gene. One of six fragments derived from the Xp21 region detected deletions in a proportion of patients with DMD and was tightly linked to the disease in family studies. Approximately 200 kb of contiguous DNA in this region was isolated by chromosome walking. The next step was to search for RNA transcripts of the suspected DMD gene. For this, non-repetitive DNA segments were screened by zoo blotting and one particular region, located 70 kb distant to the original clone, was hybridized at high stringency with chicken as well as rodent and primate DNA. Sequencing of the homologous human and mouse segments disclosed an exon bounded by splicing signals. A probe against this region detected a large RNA species in human fetal muscle which was absent from other tissues. Antisera raised against expressed portions of the gene sequence cross-react with a 400 000 kDa protein, called dystrophin, present in normal adult and fetal muscle but which is absent in DMD patients.

The identification of dystrophin by positional cloning is a vindication of the concept that uncharacterized traits can be tackled by first finding the gene and working back to the product, i.e. genetics in reverse. It also shows how conventional genetics, recombinant DNA technology, sequencing and biochemical analysis are all required for success.

Trends in positional cloning

Positional cloning now is a generally accepted method of gene identification in complex genomes, as shown by the increasing number of publications on the topic. The technique is not just restricted to human genes but has been extended to other mammals, e.g. mice (Schumacher *et al.* 1996) and to plants such as rice (Song *et al.* 1995), sugar beet (Cai *et al.* 1997) and tomato (see later). Nevertheless, positional cloning has proven to be an arduous task, with the major successes requiring the cooperative and competitive efforts of large research groups. As more and more detailed maps are constructed for organisms of interest the time-consuming steps of positional cloning should be minimized. The steps referred to here include the cloning and sub-cloning of the contiguous stretch of DNA between two genetic markers flanking the region containing the trait of interest, and the identification of all genes within this region. Even so, two problems still remain.

The first problem relates to the fact that even when the map position is known it may be difficult to identify the candidate gene. The methods described earlier for the identification of genes in large stretches of DNA are now routine in many laboratories. However, they are time consuming and do not necessarily identify all of the transcripts from a region. Until the entire genome under study is sequenced, the easiest way to find genes is to make use of ESTs provided they have been mapped to the region of interest. Finding genes in the critical region is only part of the problem, for gene-rich regions offer many candidates for analysis. Expression patterns and sequence similarity to genes of known function can aid in choosing those genes of potential interest. However, there are many examples of diseases, e.g. Huntington's chorea, with high tissue specificity which are generated by defects in genes with widespread expression. The search would be facilitated if there were easy methods to scan large stretches of DNA for mutations or sequence differences which might correspond to the different phenotypes. A number of such genome scanning methods have been developed (see p. 156).

The second problem is mapping the trait of interest to a narrow region of the chromosome, particularly if it is not associated with any obvious chromosomal abnormality. For this reason few human genes have been positionally cloned based only on point mutations. The mapping process is difficult enough in plants and animals which can be subjected to directed breeding but is extremely laborious in humans, especially for rare recessive disorders. The mapping of such traits can be facilitated by examining offspring of inbreeding, an approach termed 'homozygosity mapping' by Lander & Botstein (1987). The principle behind homozygosity mapping is that a fraction of the genome of offspring of inbreeding would be expected to be homozygous because of identity by descent. One way to simplify the search for genomic regions that are identical by descent is to use DNA pooling.

DNA pooling for linkage mapping

The concept behind the DNA pooling strategy is that identity by descent at a disease locus can be observed by pooling equal molar amounts of DNA from related affected individuals and analysing the pooled DNA sample with polymorphic STSs using PCR. The number and relative frequency of alleles at each locus can then be compared to a control DNA pool consisting of DNA from unaffected individuals. The frequency of a given allele in the population represented in the DNA pool correlates with the intensity of the allelic band on the electrophoretic gel. As DNA samples are segregated into pools on the basis of phenotype, markers in linkage disequilibrium (gametic association) with the disease locus will have different allele frequencies in the affected pool sample compared to the control pool. In contrast, unlinked markers will show similar allele frequencies, and hence band intensities, in the affected and control pools. An example of the use of DNA pooling in the identification of an autosomal recessive syndrome in humans is shown in Fig. 7.3.

Figure 7.3 The DNA pooling approach to gene mapping. (a) Pedigree of a kindred used in the DNA pooling approach. Two DNA pools were created by combining equal molar amounts of DNA from each of nine affected (solid green symbols*) and nine unaffected (open symbols*) individuals. Actual individual genotypes for a linked and unlinked marker are shown beneath each symbol. (b) A summary of the actual allele distribution and a schematic representation of the electrophoretic banding pattern to be expected with each marker when using the pooled DNA samples. S, unaffected siblings; A, affected individuals. (c) and (d) Actual data from the DNA pooling experiment illustrated in (a) showing the electrophoretic banding patterns of (c) two short tandem repeat polymorphisms that are not linked to the disease locus, and (d) two that are. Markers were amplified using the unaffected sibling (S) and the affected (A) pools as PCR templates. (Redrawn with permission from Sheffield *et al.* 1995, © *Current Opinion in Genetics and Development*.)

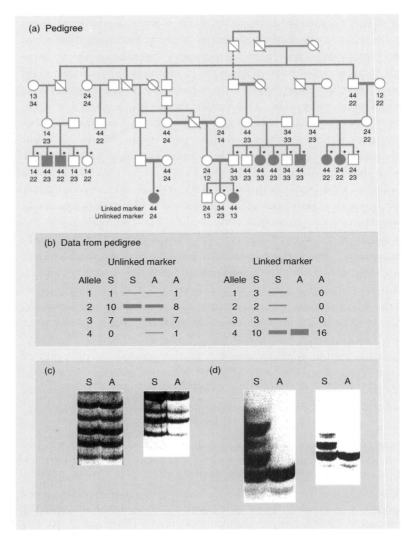

Chromosome landing for mapping plant genes

In positional cloning one finds a marker linked to a gene of interest and then walks to the gene via overlapping clones. As noted earlier (p. 64), chromosome walking is fraught with difficulties in organisms such as humans. In plants with a more complex genome organization chromosome walking is not feasible. However, the strategy of chromosome walking is based on the assumption that it is difficult to find DNA markers that are physically close to a gene of interest. In plants this need not be the case, particularly since the advent of RAPDs and AFLPs (p. 76). The plant mapping paradigm now is the isolation of one or more DNA markers at a physical distance from the targeted gene that is less than the average insert size of the genomic library being used for clone isolation. The DNA marker then is used to screen the library and isolate, or land on, the clone containing the gene of interest (for review, see Tanksley *et al.* 1995).

A good example of chromosome landing is provided by the isolation of the tomato gene *Pto* which confers resistance to the bacterial pathogen *Pseudomonas syringae* (Martin *et al.* 1993). Initially, 18 informative markers closely linked to the *Pto* gene were identified. High-resolution analysis in more than 1200 F_2 plants identified one marker that co-segregated with the gene. This marker was used to isolate YAC clones derived from the target region and genetic mapping of the isolated YAC ends identified one clone that spanned the *Pto* locus. This YAC clone was used to probe a cDNA library and all the cDNAs which hybridized were placed on a high-resolution linkage map. When this was done one group of cDNAs was identified which was closely linked to the *Pto* locus. When susceptible plants were transformed with this cDNA they acquired resistance to bacterial infection. This work exemplifies the advantages of chromosome landing in that the initial emphasis on the isolation of many closely linked DNA markers eliminated the need for chromosome walking. The development of a high-resolution linkage map by conventional genetic analysis expedited the isolation of candidate cDNAs.

The positional candidate approach to gene identification

The success of Legouis *et al.* (1991) in identifying the Kallmann syndrome gene from DNA sequence data (see p. 143) has led to an alternative approach designated the 'candidate gene' or 'positional candidate' approach (Ballabio 1993). Unlike functional and positional cloning this approach does not require the isolation of new genes but relies on the availability of information regarding function and map position from previously isolated genes (see

Chapter 5). This information could have come from either genomic sequencing projects or expressed sequence tags/cDNA fingerprints (Gerhold Caskey 1997). A good example is the human gene for X-linked glycerol kinase deficiency, where two groups spent several years isolating it by positional cloning. Meanwhile, a third group (Sargent *et al*. 1993) mapped a randomly isolated EST, homologous to *Bacillus subtilis* glycerol kinase, and located it in the human Xp21 region. After obtaining a longer cDNA sequence it was shown to be the relevant gene. Collins (1995) listed 19 human disease genes which have been identified by this method. The candidate gene approach has been used in species other than man; for example, it was used to clone 31 out of 42 mutations in the mouse (Copeland *et al*. 1993) and to clone the gene for malignant hyperthermia in pigs (Fujii *et al*. 1991). An association has also been noted between a RFLP within the gene encoding the oestrogen receptor and litter size in pigs (Rothschild *et al*. cited by Archibald 1994a).

The positional candidate approach is not without its problems as shown by Bione *et al*. (1994). They were studying the Emery–Dreifuss muscular dystrophy gene. Mutations associated with this disease were identified in a gene located in an extremely gene-dense region with numerous transcripts showing high levels of expression in muscle, the tissue affected by the disease. One gene in this small region, encoding filamin, was considered an excellent candidate for the disease gene but exhaustive analysis showed that it was not mutated in affected individuals. A comprehensive search for mutations in other genes in this critical interval led to another gene whose sequence would not lead one to expect a role in muscular dystrophy.

In future, when a new trait is assigned to a specific map position it may be possible to interrogate the genome database for that portion of the chromosome and obtain a list of other genes assigned to the same region. The features of the genes will then be compared with the features of the trait to find the most likely candidate gene. The potential gene from individuals exhibiting the trait can be screened by sequencing or hybridization to determine whether it carries sequence abnormalities.

A development from the complete sequencing of model genomes is the use of 'homology probing' to identify genes mutated in human disease (Yahraus *et al*. 1996). In this procedure, a process or pathway is identified that, when defective, results in a disease state. The genes that participate in this pathway then are identified in model organisms, and the EST database is screened for human cDNAs representing homologues of these model organism genes. These become immediate candidates for genes mutated in the human disease. This method has been used to identify human genes involved in peroxisome biogenesis.

Studying complex and quantitative traits

The term 'complex trait' refers to any phenotype that does not exhibit classic Mendelian recessive or dominant inheritance attributable to a single gene locus. Most, but not all, complex traits can be explained by polygenic inheritance. That is, these traits require the simultaneous presence of mutations in multiple genes. Polygenic traits may be classified (Lander & Schork 1994) as discrete traits, measured by specific outcome (e.g. development of diabetes or death from myocardial infarction), or quantitative traits measured by a continuous variable, (e.g. grain yield, body weight).

Many important biological characteristics are inherited quantitatively but, because these effects have not generally been resolvable individually, quantitative geneticists have dealt largely with characterization of these factors using biometrical procedures. However, many issues in quantitative genetics and evolution are difficult to address without additional information about the genes that underlie continuous variation. The identification of such *quantitative trait loci* (QTLs) has become possible with the advent of RFLPs as genetic markers and the increasing availability of complete RFLP maps in many organisms.

The basis of all QTL detection, regardless of the species to which it is applied, is the identification of association between genetically determined phenotypes and specific genetic markers. However, there is a special problem in mapping QTLs and other complex trait genes and that is penetrance, i.e. the degree to which the transmission of a gene results in the expression of a trait. For a single gene trait, biological or environmental limitation accounts for penetrance, but in a multigene trait the genetic context is important. Hence the consequences of inheriting one gene rely heavily on the co-inheritance of others. For this reason careful thought needs to be given to the strains selected for analysis and the genetic crosses made.

In inbreeding populations, i.e. crops and laboratory and domestic animals, the kinds of cross used are shown in Fig. 7.4. Populations typically are various crosses between inbred strains, usually the first generation backcross (N_2) or intercross (F_2). Higher generation crosses (N_3), panels of recombinant inbred strains (RI), recombinant congenic strains (RC), or inbred strains themselves also may be used. The chromosomal content in these panels is the heart of the study. They define which alleles are inherited in individuals, so that chromosomal associations can be made, and provide genetic recombination information so that location within a chromosome can be deduced. F_1 hybrids are genetically identical to each other but individuals in subsequent generations are not. Backcross progeny reveal recombinations on only one homologue, the one inherited from the F_1 parent, but intercross progeny reveal them on both homologues. RI and RC strains also harbour recombinations but, unlike backcross and intercross progeny, these are homozygous at all loci as a result of inbreeding. A congenic strain has only one chromosomal region that distinguishes it from a parental strain. Because they have an unchanging genotype, RI, RC and congenic strains offer an elegant way of discriminating between the role of the environment and of genetic factors in the expression of a phenotype.

There are a number of approaches to mapping QTLs. All involve arranging a cross between two inbred strains differing substantially in a quantitative trait. Segregating progeny, of the type shown in Fig. 7.4 and described above, are scored both for the trait and for a number of genetic markers. In the case of plants the markers would be RFLPs or AFLPs and in the case of animals would be microsatellites (polymorphic STSs).

In the method of Edwards *et al.* (1987), linear regression is used

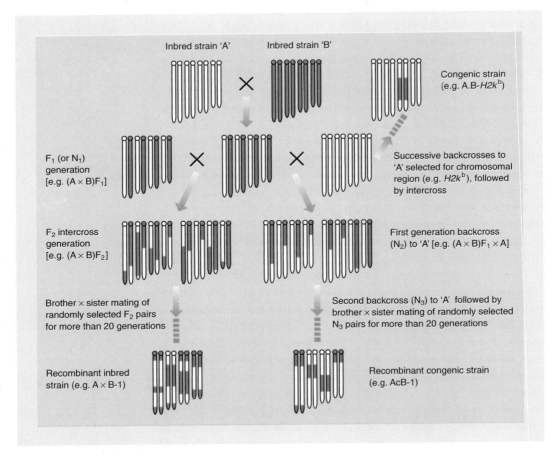

Figure 7.4 Crosses used in the analysis of complex traits. (Adapted with permission from Frankel 1995.)

to examine the relationship between the performance for the quantitative trait and the genotypes at the marker locus. If there is a statistically significant association between the trait performance and the marker locus gene types (Fig. 7.5) it is inferred that a QTL is located near the marker locus.

Lander and Botstein (1989) have pointed out that there are a number of disadvantages with the above method. In particular, it cannot distinguish between tight linkage to a QTL with small effect and loose linkage to a QTL with large effect. This is not a problem with the second method, called *interval mapping*, which is an adaptation of the logarithm of odds (LOD) score analysis used in human genetics.

A LOD score is a statistical assessment of evidence for, in this case, linkage between a QTL and a marker. This is the logarithm of the ratio of two pedigree data likelihoods, the one obtained assuming no linkage divided by the one obtained with no linkage. A LOD score of 3 or higher (NB this is a logarithmic value) is usually accepted as demonstrating that linkage is present and in most situations it roughly corresponds to the 5% level of significance used in conventional statistical tests. However, lower LOD scores can still indicate linkage. For interval mapping one calculates the LOD score for different marker loci along the chromosome, e.g. from marker loci a to e in Fig. 7.6. On the basis of the size of the genome and the number of marker loci analysed a threshold value or significance level is determined. In the example shown in Fig. 7.6 this threshold value has been set at 2.5. Where the LOD score exceeds the threshold a QTL is likely. Haley *et al.* (1994) have adapted the interval mapping approach for mapping QTLs in outbreeding populations.

The above methods have been used to map QTLs in both crops and livestock animals; for example, Paterson *et al.* (1988) mapped 15 QTLs in the tomato affecting fruit mass, concentration of

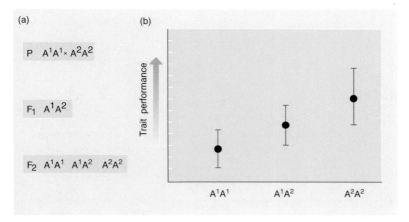

(a)

P $A^1A^1 \times A^2A^2$

F_1 A^1A^2

F_2 A^1A^1 A^1A^2 A^2A^2

(b)

Trait performance

A^1A^1 A^1A^2 A^2A^2

Figure 7.5 Relationship between the performance for a quantitative trait and genotypes at a marker locus for an F_2 interbreeding population. (a) Genetic composition of the F_2 population sampled. (b) Quantitative performance of different types of F_2 genotypes.

soluble solids and fruit pH. Andersson *et al.* (1994) generated an F_2 intercross between wild boars and domestic pigs. Two hundred offspring were genotyped for 105 polymorphic markers. In addition, growth rates were recorded from birth until they weighed 70 kg, and after slaughter fat levels and intestinal length were noted. Fatness, whether measured as a percentage of fat in the abdominal cavity or backfat thickness, mapped to the proximal end of chromosome 4. The QTL for intestinal length and growth rate were located distal to the fatness QTL.

The analysis of complex traits in humans demands a different approach. One has to start with the analysis of families of affected individuals. Affected members of a family will share the genes causing the disease and this sharing will manifest itself as an excess of sharing over the 50% expected at random in siblings. Disease genes can be identified by co-segregation with linked genetic markers which will show identity by descent in affected individuals. The markers used must be close enough that recombination with the disease locus is unlikely and sharing among individuals can only be tested if the markers are polymorphic. Microsatellite markers are ideal for this purpose as they are highly polymorphic and sufficient numbers of them have been positioned throughout the genome. Using such microsatellite markers significant progress has been made in understanding the genetic basis of type 1 diabetes (Cordell & Todd 1995), Alzheimer's disease (Pericak-Vance & Haines 1995), asthma (Daniels *et al.* 1996) and stroke (Rubattu *et al.* 1996).

The development of methods for mapping and characterizing QTLs can be used to study evolution at the molecular level. A good example is provided by maize (Indian corn) and its relatives, the teosintes. Although they differ in both ear (corn cob) morphology and plant growth form (Fig. 7.7), it is believed that maize is a domestic form of teosinte. Indeed, a small number of genes selected by the prehistoric peoples of Mexico may have trans-

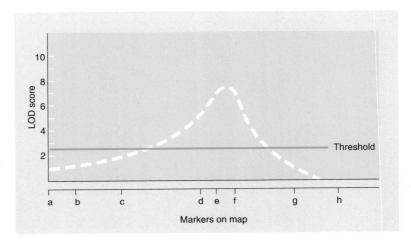

Figure 7.6 Interval mapping to determine whether a QTL is located near a particular genetic marker. The distances between the genetic markers represent genetic distance.

formed teosinte into maize within the past 10 000 years. Morpho-logically there are five points of difference between the two plants and analysis by classical genetics has suggested that five major and independently inherited gene differences distinguish the two species. Doebley (1992) has confirmed this by molecular methods and has shown that the major differences map to five restricted regions of the genome, each on a different chromosome.

The reader with a special interest in the use of molecular methods for dissecting QTLs and other complex traits should consult the December 1995 issue of *Trends in Genetics* (vol. 11, no. 12) which was devoted to this subject. The organisms covered include maize, rice, mice, various livestock, *Drosophila* (genetic variation in bristle content) and humans. Various specialist methodologies also are discussed.

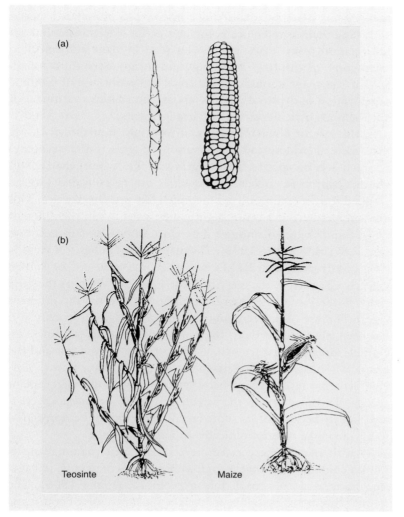

Figure 7.7 Comparison of (a) ear structure, and (b) plant morphology for maize and its ancestor teosinte. (Adapted and redrawn with permission from Doebley 1992.)

Comparative genomic mapping

In the pig study noted above (p. 152), the QTLs were mapped to a region of the genome over 15 cM in length. This equates to almost 15 Mb of DNA sequence and hence positional cloning of the genes is not feasible unless this distance can be reduced by generating a more dense map. Without this, the positional candidate-gene approach is more likely to be successful (Archibald 1994a). However, this approach depends on the linkage map being well populated with known genes and for most organisms this is not yet the case. One solution may be to make use of comparative genomic mapping.

Using the molecular tools discussed in Chapter 4, detailed genetic maps are being constructed in a wide range of plants and animals. Initially these maps were compiled to provide a resource for genetic analysis in the species selected. More recently it has been realized that comparison of maps from related species has two benefits. First, by identifying conserved loci in related species and using them as reference points it is possible to transfer linkage information from 'map-rich' species (e.g. humans and mice) to 'map-poor' species (e.g. cows, pigs and sheep). Second, comparison of the maps should help dissect the evolution of genome organization and provides clues about the adaptive rationale, if any, behind particular structural arrangements.

In the case of flowering plants, it now has been shown clearly that there is widespread conservation of gene order (synteny) between related species. In the grasses (rice, wheat, maize, sorghum, sugar cane, millet) the genomes can be considered as descended from a common genome much like present day rice. This genome has been cut up in slightly different ways in the different crops and distributed amongst different numbers of chromosomes (Fig. 7.8) (Moore *et al.* 1995). The maps shown in Fig. 7.8 initially were constructed with RFLPs but when the locations of known genes were added to the maps they too lined up. Whereas the gene order is the same, the intergenic distances vary in the different cereals because of different amounts of repeat DNA. Synteny also has been observed between the potato and tomato genomes, including disease-resistance genes, and between *Arabidopsis* and the brassicas (Shields 1996). Paterson *et al.* (1996) have now discovered syntenic relationships between more distantly related plant species. This may permit the wealth of data on *Arabidopsis*, generated largely because of its short breeding period, to be transferred to a wide range of agronomically important crops.

As with plants, there is ample evidence that mammalian genomes contain significant recognizable regions of genetic linkages that have been conserved during evolution (Eppig 1996). The human/mouse comparative map is the most extensive because the

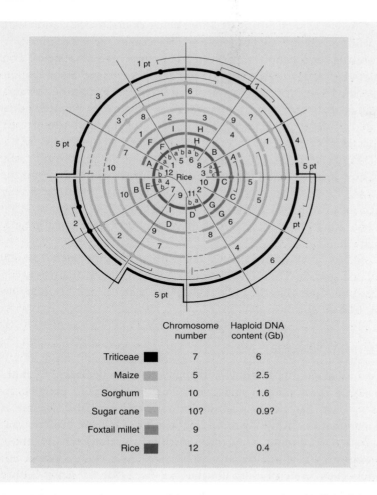

	Chromosome number	Haploid DNA content (Gb)
Triticeae ■	7	6
Maize	5	2.5
Sorghum	10	1.6
Sugar cane	10?	0.9?
Foxtail millet ■	9	
Rice ■	12	0.4

Figure 7.8 Alignment of the genomes of six major grass crop species with 19 rice linkage segments, whose order reflects the circularized ancestral grass genome. The data have been redrawn as a series of rice linkage segments (defined by radiating lines) formed into chromosomes (broken colour-coded and numbered lines). The thin dashed lines correspond to the duplicated segments. Inversions of sets of sequences within a linkage segment (such as the inversion of segments 3a and 3b in maize chromosome 5) are not shown. Linkage segments forming parts (pt) of Triticeae chromosome 5 are shown as a series of segments connected by coloured lines. The alignment is based on the genetic map of the D genome of wheat. The dark green line indicates the duplicated segments shown as blocks 11b, and 12b. Chromosomes formed by the insertion of one segment into another are shown by black lines with arrows indicating the direction and point of insertion. The points of chromosome breakage involved with insertion events are indicated by black bisected circles. (Redrawn with permission from Moore *et al.* 1995, © *Current Biology*.)

genetics of these two mammalian species are still the most widely and best studied. As of late 1996 this map comprised 105 conserved autosomal segments and eight conserved X chromosome segments. The existence of these comparative maps is facilitating the rapid increase in loci being placed on the genetic maps of other mammalian species.

Chromosome painting by fluorescence *in situ* hybridization with chromosome-specific libraries provides a quick, generalized way to compare the location of conserved segments between species. This method, termed ZOO-FISH utilizes composite chromosome-specific DNA probes from one species to hybridize to the chromosomes of another species. Major regions of conserved synteny thus can be identified between the 'probe' species and the 'target' species. The reported lower limit of resolution is 7 Mb. ZOO-FISH has been used to show synteny between humans and cattle (Chowdhary *et al.* 1996) and between humans and pigs (Fronicke *et al.* 1996). The method has great potential for determining genomic similarity between diverse mammalian genomes, even when some of these genomes have a poorly developed genetic map.

As was observed in plants, synteny can extend far beyond closely related species. Thus Trower *et al.* (1996) found synteny between a group of genes in the puffer fish (*Fugu rubripes*) and homologues on human chromosome 14.

Genome scanning methods

Positional cloning is a powerful approach to the isolation of genes which is applicable, in principle, to any organism. In practice, its actual use has been much more restricted because success is dependent on the ability to find tightly linked genetic markers near a locus of interest. Thus the method is restricted to those organisms for which dense genetic maps have been constructed. Even then, determining linkage can be a lot of work. Other methods of gene isolation depend on some knowledge of the function of the gene product and, again, this may not be available. Ideally, what is required are methods that allow an entire genome to be scanned at one time for sequences linked to trait loci. A number of these, termed *genome scanning methods* (Brown 1994), have been developed and are described below. Interestingly, with them, classical genetic techniques are as important as the molecular methods used. It is worth noting that all of the methods described below may be superseded by the use of hybridization to high-density oligonucleotide arrays on microchips as a result of the success of Chee *et al.* (1996) in rapidly detecting human mitochondrial variants (p. 124).

GENETICALLY DIRECTED REPRESENTATIONAL DIFFERENCE ANALYSIS (GDRDA)

GDRDA is based on a subtractive technique, known as representational difference analysis (RDA), for identifying differences between two DNA samples (Lisitsyn *et al.* 1993). The two DNA samples are known as tester and driver. Specifically, RDA is designed to clone restriction fragments that can be amplified by PCR from tester but not driver. This may be because the corresponding sequence is completely absent from the driver due to a homozygous deletion or because it is contained in a small restriction fragment in the tester but in a large and, therefore poorly amplifiable, restriction fragment in the driver. Thus RDA exploits size differences in DNA from driver and tester.

The principle of RDA is shown in Fig. 7.9. Restriction-enzyme-digested genomic DNA fragments are first ligated to adaptor

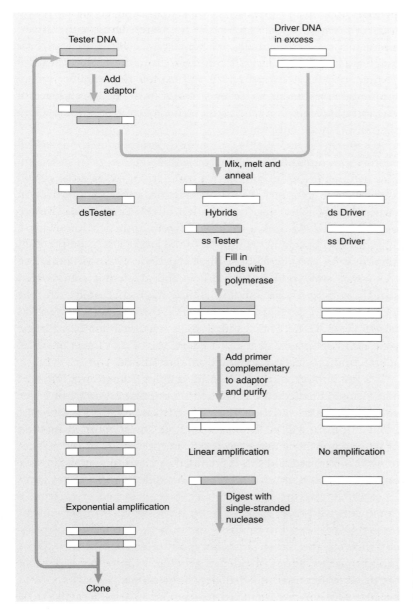

Figure 7.9 The principle of representational difference analysis (RDA). See text for details.

oligonucleotides to allow their amplification using a cognate primer in PCR. Because of the biased amplification of fragments in the initial complex pool principally favouring fragments of smaller size, the pool of products (an 'amplicon') obtained after PCR amplification of the initial sample constitutes a 'representation' of the genome. The complexity of this may be substantially less than that of the original genomic DNA. Ligation of a second adaptor oligonucleotide to the 3′ ends of the fragments in one of the representations (the 'tester' amplicon) provides a means for selectively amplifying duplex DNA molecules, both strands of which derive from the 'tester' amplicon. Thus, after denaturation and hybridization in the presence of an excess of a second amplicon ('driver'), sequences represented in the tester, but not the driver, amplicon are preferentially recovered in a form that can be amplified by subsequent PCR. Successive rounds of selection allow further subtractive hybridization and exploit the amplification of the initial enrichment factor that results from the dependence of the annealing rate of each fragment on its relative abundance in the pool (kinetic enrichment).

Rosenberg *et al.* (1994) described another method, RFLP subtraction, designed to purify restriction fragments from a complex genome if they do not have a counterpart of the same size in a competing genome. RFLP subtraction uses gel purification rather than PCR for size fractionation of DNA fragments. Rosenberg *et al.* (1994) took a slightly different approach for selective amplification of fragments unique to the target DNA. Before PCR amplification they performed three repetitive cycles of annealing the target with biotinylated driver and depleting biotinylated sequences with avidin binding. Once fragments unique to the target had been concentrated sufficiently, the method involved was identical to RDA. These added steps appeared to increase the power and specificity of the procedure: 21 of 22 clones of RFLP subtraction products represented unique RFLPs.

It is not enough to apply RDA to samples from a single affected and a single unaffected individual in a population or family. The abundant genetic variation among even close relatives means that polymorphisms will be found throughout the genome. A method of finding polymorphisms specifically in the vicinity of the gene of interest is required and this is provided by classical genetic crosses. Suppose A and B are two inbred strains differing at a target locus L, strain A carrying a mutant recessive allele and B carrying a dominant wild-type allele. To create a driver sample, an F_2 intercross between the strains is performed, a collection of progeny showing the recessive phenotype is selected, and their DNA mixed together. From Mendelian genetics it can be predicted that the driver contains (i) no B alleles in the immediate vicinity of L, because progeny were selected for the recessive phenotype; (ii) a deficit of B alleles in a somewhat larger region around L, owing to linkage to L; and (iii) roughly equal proportions of A and B alleles elsewhere in the genome. If RDA is performed with this driver and DNA from strain B as tester, then B alleles should be subtracted everywhere in the genome except in a region around L.

The targeting of the method can be improved further if locus L has been genetically mapped and two flanking markers (X and Y) identified. For the driver $k/2$ progeny are selected in which cross-over occurred between L and X and $k/2$ progeny in which the cross-over occurred between L and Y (where k is the total number of progeny). This refinement of the method should allow targeting of very small intervals of DNA and Lisitsyn *et al.* (1994) confirmed this by generating probes mapping close to two target mouse loci. GDRDA is applicable to more than just F_2 intercrosses between inbred strains. It can be applied to backcrosses between inbred strains, two-generation families in an outbred population (e.g. humans where inbred lines are not readily available) and half-sib mating schemes (common in livestock breeding). The technique has been used to detect genetic lesions in cancer. In this case, the tester DNA sample is prepared from normal DNA and the driver DNA sample from pure tumour DNA derived from the same patient (Lisitsyn 1995). An alternative approach is to use tumour DNA for preparation of the tester sample and normal DNA for preparation of the driver sample. This experimental approach generates sequences that detect high-level amplifications in tumours.

To date, the use of RDA and RFLP subtraction in plant studies has not been reported. However, in the technique called chromosome landing (p. 147), RAPDs have been used in a similar way (for review see Tanksley *et al.* 1995).

GENOME MISMATCH SCANNING (GMS)

Whereas GDRDA focuses on dissimilarities between two genomes, GMS seeks to identify large regions of sequence identity between two individuals (Nelson *et al.* 1993). The rationale is that because the rate of natural polymorphism in a population is substantial, identical sequences must represent genomic regions of 'identity by descent'. The principle of GMS is shown in Fig. 7.10. High molecular weight DNA is prepared from two related individuals. Both genomic DNAs are digested with the same restriction endonuclease which must leave protruding 3' ends. The 3' protrusions serve to protect these ends from digestion with exonuclease III in later steps. Also, the DNA fragments need to be long enough so that many fragments are likely to contain both a GATC sequence and at least one base that differs between unrelated allelic fragments. To differentiate the two DNAs, one individual's DNA is fully methylated at all GATC sites with the *E. coli* Dam methylase. The other individual's DNA remains unmethylated at all GATC sites. The two restriction-digested genomic DNA

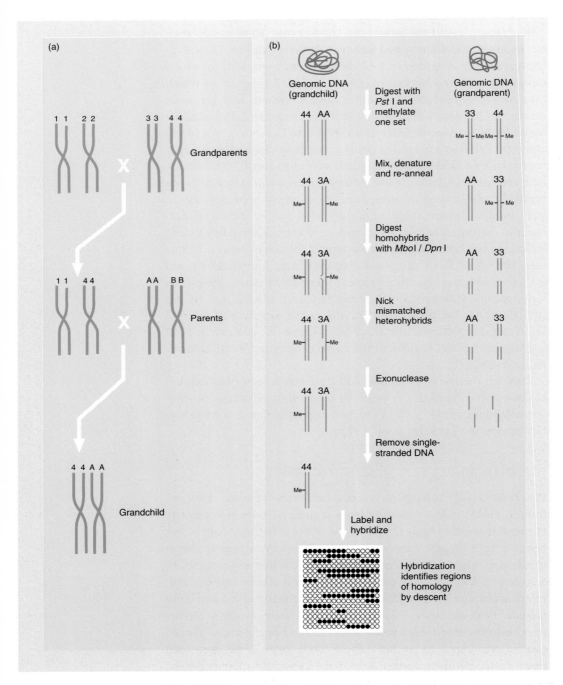

Figure 7.10 The principle of genome mismatch scanning (GMS). (a) The relationship between the chromosomes of the grandchild and grandparents. (b) Experimental method to generate regions of homology. See text for details.

samples are then mixed in equimolar ratios, denatured and allowed to reanneal. Approximately 50% of the reannealed fragments are heterohybrids and are, therefore, hemimethylated (methylated on only one strand) at each GATC sequence. 'Homohybrids', formed from two strands of the same individual's DNA, are either methylated on both strands or not methylated at all. Digestion of the reannealed DNA with both *Dpn*I and *Mbo*I, which cut at fully methylated and unmethylated GATC sites respectively, results in cleavage of the homohybrids to yield smaller duplexes with either blunt ends or 5′ protruding ends. The heterohybrids are resistant to cleavage by both *Dpn*I and *Mbo*I. This step therefore distinguishes between homohybrids and heterohybrids.

Sensitive distinction between mismatch-free heterohybrids and those with base mismatches is achieved using the methyl-directed mismatch repair system of *E. coli*. Three proteins, MutL, MutS and MutH, are sufficient to recognize seven of the possible eight single-base-pair mismatches in duplex DNA. They are able to introduce a single strand nick on the unmethylated strand at GATC sites specifically in the mismatch containing duplexes. The heterohybrids are incubated with MutL, MutS and MutH. Only those that are mismatch-free, and therefore likely to be from regions of 'identity by descent', escape nicking.

All DNA molecules, except the mismatch-free heterohybrids, can be degraded with ExoIII, a 3′ to 5′ exonuclease specific for double-stranded DNA (dsDNA). ExoIII can initiate digestion at a nick and degrade one strand to yield a single-stranded DNA (ssDNA) gap. ExoIII can also initiate at the blunt or 5′ protruding ends produced by cleavage with *Dpn*I or *Mbo*I, respectively, to degrade the homohybrids to ssDNA. The heterohybrids have only 3′ protruding ends which resist degradation by exonuclease III. Thus, only DNA duplexes which are both heterohybrids and free of mismatches escape digestion. Molecules containing single-stranded DNA are removed by chromatography. The purified heterohybrids are labelled and used to probe an ordered array of DNAs representing the entire genome. Regions of hybridization represent regions of identity by descent.

Nelson *et al.* (1993) have tested GMS experimentally using *Saccharomyces cerevisiae* as a model system. In their test, two parental yeast strains, A and B, were mated and the resulting diploid was sporulated to produce haploid progeny. Progeny were compared to each parent using GMS, with the resulting DNA being hybridized to a collection of consecutive phage clones covering the length of chromosome 5. As expected, the chromosome can be divided into blocks that are identical to either the A or the B parent, with the transitions corresponding to the site of cross-over (Fig. 7.11). The results were unambiguous for 20 of 31 phage clones. The remaining 11 clones were uninformative for

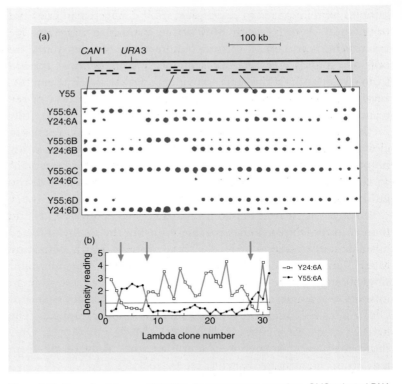

Figure 7.11 Example of the use of GMS in *Saccharomyces cerevisae*. GMS selected DNA from parent Y55 and Y24/spore clone (6A, B, C and D) pairs were hybridized onto an ordered array of 31 λ clones representing 79% coverage chromosome 5. (a) The intervals spanned by each λ clone are indicated by bars above the dot blot. *CAN*1 and *URA*3 are two yeast genes mapped to chromosome 5. Row 1, Y55, shows the result of hybridization using unselected DNA from the Y55 parent as a probe to control for variability in the amount of DNA immobilized at each dot. Rows 2–9 show results of hybridization using GMS-selected DNA from the indicated parent : spore clone pair. (b) Normalized densitometer reading of the hybridization signals from the Y24 : 6A and Y55 : 6A pairwise comparisons. Meiotic recombination breakpoints are indicated by the arrows. (Redrawn with permission from Nelson *et al.* 1993, courtesy of *Nature Genetics*.)

this GMS-based linkage analysis because they gave consistently positive or consistently negative results, most probably because of technical artefacts. If GMS can be applied satisfactorily to much larger genomes then it could facilitate the study of polygenic traits. A search would be made for regions of 'identity by descent' in pairs of affected cousins or in a collection of affected individuals from an isolated subpopulation.

SCREENING BY ENZYME MISMATCH CLEAVAGE (EMC)

The two genome scanning methods described above are not ideal. Although RDA has been shown to identify a specific locus in highly inbred mouse strains it really is not applicable to highly outbred populations such as humans. GMS is effective in yeast but has not yet been shown to work on mammalian genomes which are much more complex and contain repetitive sequences. In addition, the unique sequences in the human genome are less variable between individuals than yeast sequences. By contrast, EMC has been developed for identifying mutations in human genes (Mashal *et al*. 1995; Youil *et al*. 1995).

Bacteriophage resolvases are enzymes whose function *in vivo* is to cleave branched DNA. They have the property of recognizing mismatched bases in double-stranded DNA and cutting the DNA at the mismatch. EMC takes advantage of this characteristic to detect individuals who are heterozygous at a given site. Radiolabelled DNA is cleaved by the resolvase at the site of mismatch in heteroduplex DNA and digestion is monitored on a gel (Fig. 7.12). Thus, both the presence and the estimated position of an alteration are revealed. In an analysis of all possible mismatch types as well as several deletion mutations, an overall sensitivity of mutation detection of 94% was achieved.

A review of the progress in developing human genome scanning methods has been presented by Cotton (1997).

Understanding the phenotype

Once a gene has been isolated it is essential to be sure that it is the correct one. Following on from this there usually will be a need to understand how the gene functions or why mutant genes produce the effects that they do. In many experimental systems these different needs can be met simultaneously. For example, in the case of the tomato gene *Pto* described earlier (p. 147), transfection of susceptible plants with the cloned gene resulted in resistance to bacterial infection. The availability of a cloned gene, which can be sequenced, and the availability of resistant and susceptible plants, which can be probed at the genomic level, means that the phenotype can be dissected. The same is true for most genes in most other systems. There is one exception and that is humans. In this case the best that can be done is to try to model the disease state in another mammal and the one chosen is the mouse.

The mouse is a near-perfect experimental system for creating models of human genetic diseases. The genome size and number of genes are similar between the human and mouse as are patterns of development. Mice have relatively short gestation periods, brief time of development to sexual maturity and large litter sizes. Finally, it is the only one where currently it is possible to perform germline knockouts in embryonic stem cells. The repertoire in mice now includes deletion of specific genes, introduction of particular mutations, overexpression, gene dosage modifications and chromosome dosage modifications (Lamb & Gearhart 1995; Ramirez-Solis *et al*. 1995; Bradley & Liu 1996). Table 7.3 lists some examples of these model systems. However, modifying an orthologous gene in mice does not always result in the creation of the same syndrome as seen in humans and there are many reasons for this (Wynshaw-Boris 1996).

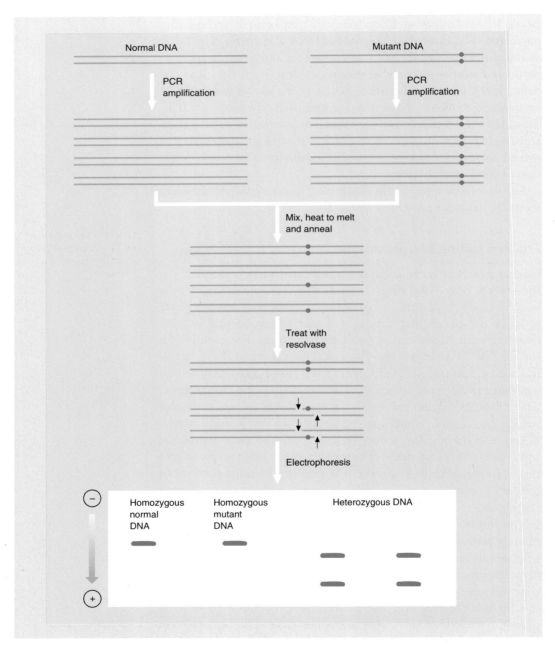

Figure 7.12 The principle of enzyme mismatch cleavage. See text for details.

In principle, it is possible to construct mice with mutations in any cloned gene. However, for genes cloned initially as cDNAs, one must isolate and characterize genomic clones, construct targeting vectors, screen embryonic stem cell clones to identify those in which the genes have been disrupted and then develop chimaeric mice. Although many genes have been disrupted in this manner the process is slow and laborious. Consequently, gene-trapping strategies have been developed to disrupt genes expressed in mouse embryonic stem cells. A promoter-less selectable marker is introduced into the cells and clones expressing the gene are selected when the targeting vector inserts into, and disrupts, expressed cellular genes (Hicks *et al*. 1997).

It should be realized that the full understanding of the role of a gene comes about when one knows how the gene product interacts with other genes or gene products, i.e. identity of gene function is not enough. Little attention has been paid to this aspect of genome analysis but Fromont-Racine *et al*. (1997) have suggested a way in which such information might be obtained in yeast.

Genetic alteration	Human disease equivalent
Introduction of mutant collagen gene into wild-type mice	Osteogenesis
Inactivation of mouse gene coding hypoxanthine-guanine phosphoribosyl transferase (HPRT)	HPRT deficiency
Mutation of locus for X-linked muscular dystrophy	X-linked muscular dystrophy
Introduction of activated human *ras* and *c-myc* oncogenes	Induction of malignancy
Introduction of mutant (Z) allele of human α_1 antitrypsin gene	Neonatal hepatitis
Introduction of HIV *tat* gene	Kaposi's sarcoma
Overproduction of atrial natriuretic factor	Chronic hypotension
Introduction of rat angiotensinogen gene	Hypertension
Constitutively active tyrosine kinase	Cardiac hypertrophy
Overexpression of amyloid precursor protein	Alzheimer's disease
Trisomy 16	Alzheimer's disease
Expression of Simian cholesteryl ester transfer protein	Atherosclerosis

Table 7.3 Human disease equivalents derived from genetic alterations in the mouse

References

Abidi F., Wada M., Little R.D. and Schlessinger D. (1990) Yeast artificial chromosomes containing human Xq24–Xq28 DNA: library construction and representation of probe sequences. *Genomics* 7, 363–376.

Adams M.D. and 12 others (1991) Complementary DNA sequencing: expressed sequence tags and human genome project. *Science* **252**, 1651–1656.

Adams M.D. and 84 others (1995) Initial assessment of human gene diversity and expression patterns based upon 83 million nucleotides of cDNA sequence. *Nature* **377**, S3–S173.

Adams M.D., Dubnick M., Kerlavage A.R., Morenoo R., Kelley J.M., Utterback T.R., Nagle J.W., Fields C. and Venter J.C. (1992) Sequence identification of 2,375 human brain genes. *Nature* **355**, 632–634.

Adams M.D., Kerlavage A.R., Fields C. and Venter J.C. (1993a) 3,400 new expressed sequence tags identify diversity of transcripts in human brain. *Nature Genetics* **4**, 256–267.

Adams M.D., Soares M.B., Kerlavage A.R., Fields C. and Venter J.C. (1993b) Rapid cDNA sequencing (expressed sequence tags) from a directionally cloned human infant brain cDNA library. *Nature Genetics* **4**, 373–380.

Ahringer J. (1997) Turn to the worm. *Current Opinion in Genetics and Development* 7, 410–415.

Alford R.L. and Caskey C.T. (1994) DNA analysis in forensics, disease and animal/plant identification. *Current Opinion in Biotechnology* **5**, 29–33.

Altschul S.F., Boguski M.S., Gish W. and Woolton J.C. (1994) Issues in searching molecular databases. *Nature Genetics* **6**, 119–129.

Anand R., Riley J.H., Butler R., Smith J.C. and Markham A.F. (1990) A 3.5 genome equivalent multi-access YAC library: construction, characterisation, screening and storage. *Nucleic Acids Research* **18**, 1951–1956.

Andersson L. and 11 others (1994) Genetic mapping of quantitative trait loci for growth and fatness in pigs. *Science* **263**, 1771–1774.

Archibald A.L. (1994a) Fat pigs can blame their genes. *Current Biology* **4**, 728–730.

Archibald A.L. (1994b) Mapping of the pig genome. *Current Opinion in Genetics and Development* **4**, 395–400.

Archibald A.L. and 52 others (1995) The PiGMaP consortium linkage map of the pig (*Sus scrofa*). *Mammalian Genome* **6**, 157–175.

Baer R. and 11 others (1984) DNA sequence and expression of the B95.8 Epstein–Barr virus genome. *Nature* **310**, 207–211.

References

Bains W. and Smith G.C. (1988) A novel method for nucleic acid sequence determination. *Journal of Theoretical Biology* **135**, 303–307.

Ballabio A. (1993) The rise and fall of positional cloning. *Nature Genetics* **3**, 277–279.

Bankier A.T. and Barrell B.G. (1983) Shotgun DNA sequencing. *Technol. Nucleic Acid Biochemistry* **B5**, 1–34.

Barbour A.G. (1993) Linear DNA of *Borrelia* species and antigenic variation. *Trends in Microbiology* **1**, 236–239.

Barendse W. and 22 others (1994) A genetic linkage map of the bovine genome. *Nature Genetics* **6**, 227–235.

Barrell B. (1991) DNA sequencing: present limitations and prospects for the future. *FASEB Journal* **5**, 40–45.

Barrett J.H. (1992) Genetic mapping based on radiation hybrid data. *Genomics* **13**, 95–103.

Belfort M. (1989) Bacteriophage introns: parasites within parasites? *Trends in Genetics* **5**, 209–213.

Bellanné-Chantelot C. and 21 others (1992) Mapping the whole human genome by fingerprinting yeast artificial chromosomes. *Cell* **70**, 1059–1068.

Bender W., Spierer P. and Hogness D.S. (1983) Chromosome walking and jumping to isolate DNA from the *Ace* and *rosy* loci and the bithorax complex in *Drosophila melanogaster*. *Journal of Molecular Biology* **168**, 17–33.

Bennetzen J.L. (1996) The contributions of retroelements to plant genome organization, function and evolution. *Trends in Microbiology* **4**, 347–353.

Benson D.A., Boguski M., Lipman D.J. and Ostell J. (1996) GenBank. *Nucleic Acids Research* **24**, 1–5.

Benson D.A., Boguski M., Lipman D.J. and Ostell J. (1997) GenBank. *Nucleic Acids Research* **25**, 1–6.

Bentley D.R. and Dunham I. (1995) Mapping human chromosomes. *Current Opinion in Genetics and Development* **5**, 328–334.

Bione S., Maestrini E., Rivella S., Mancini M., Regis S., Romeo G. and Toniolo D. (1994) Identification of a novel X-linked gene responsible for Emery–Dreifuss muscular dystrophy. *Nature Genetics* **8**, 323–327.

Bird A. (1995) Gene number, noise reduction and biological complexity. *Trends in Genetics* **11**, 94–99.

Blattner F. and 16 others (1997) The complete genome sequence of *Escherichia coli* K-12. *Science* **277**, 1453–1462.

Boguski M.S. (1995) The turning point in genome research. *Trends in Biochemical Sciences* **20**, 295–296.

Boguski M.S., Lowe T.M.J. and Tolstoshev G.M. (1993) dbEST-database for 'expressed sequence tags'. *Nature Genetics* **4**, 332–333.

Bork P. (1996) Go hunting in sequence databases but watch out for the traps. *Trends in Genetics* **12**, 425–427.

Botstein D., White R.L., Skolnick M. and Davis R.W. (1980) Construction of a genetic linkage map in man using restriction fragment length polymorphisms. *American Journal of Human Genetics* **32**, 314–331.

Bradley A. and Liu P. (1996) Target practice in transgenics. *Nature Genetics* **14**, 121–123.

Bradshaw H.D., Wilbert S.M., Otto K.G. and Schemske D.W. (1995)

Genetic mapping of floral traits associated with reproductive isolation in monkeyflowers (*Mimulus*). *Nature* **376**, 762–765.

Branscomb E., Slezak T., Pae E., Gales D., Carrano A.V. and Waterman M. (1990) Optimizing restriction fragment fingerprinting methods for ordering large genomic libraries. *Genomics* **8**, 351–366.

Breatnach R., Mandel J.L. and Chambon P. (1977) Ovalbumin gene is split in chicken DNA. *Nature* **270**, 314–319.

Brenner S.E. (1995) Network sequence retrieval. *Trends in Genetics* **11**, 247–248.

Brenner S., Elgar G., Sandford R., Macrae A., Venkatesh B. and Aparicio S. (1993) Characterisation of the pufferfish (*Fugu*) genome as a compact model vertebrate genome. *Nature* **366**, 265–268.

Britten R.J. and Kohne D.E. (1968) Repeated sequences in DNA. *Science* **161**, 529–540.

Brown P.O. (1994) Genome scanning methods. *Current Opinion in Genetics and Development* **4**, 366–373.

Brown W.R.A. (1992) Mammalian artificial chromosomes. *Current Opinion in Genetics and Development* **2**, 479–486.

Buckler A.J., Chang D.D., Graw S.L., Brook D., Haber D.A., Sharp P.A. and Houseman D.E. (1991) Exon amplification: a strategy to isolate mammalian genes based on RNA splicing. *Proceedings of the National Acadmy of Sciences. USA* **88**, 4005–4009.

Bult C.J. and 39 others (1996) Complete genome sequence of the methanogenic archaeon, *Methanococcus jannaschii*. *Science* **273**, 1058–1073.

Burke D.T., Carle G.F. and Olson M.V. (1987) Cloning of large segments of exogenous DNA into yeast by means of artificial chromosome vectors. *Science* **236**, 806–813.

Cai D. and 12 others (1997) Positional cloning of a gene for nematode resistance in sugar beet. *Science* **275**, 832–838.

Cantor C.R., Smith C.L. and Mathew M.K. (1988) Pulsed-field gel electrophoresis of very large DNA molecules. *Annual Review of Biophysical Chemistry* **17**, 287–304.

Casari G., De Daruvar A., Sander C. and Schneider R. (1996) Bioinformatics and the discovery of gene function. *Trends in Genetics* **12**, 244–245.

Cellier M., Belouchi A. and Gros P. (1996) Resistance to intracellular infections: genomic analysis of *Nramp*. *Trends in Genetics* **12**, 201–203.

Chakrabarti D., Reddy G.R., Dame J.B., Almira A.C., Laipis P.J., Feri R.J., Yang T.P., Rowe T.C. and Schuster S.M. (1994) Analysis of expressed sequence tags from *Plasmodium falciparum*. *Molecular Biochemical Parasitology* **66**, 97–104.

Chee M. and 9 others (1996) Accessing genetic information with high-density arrays. *Science* **274**, 610–614.

Chen C.W. (1996) Complications and implications of linear bacterial chromosomes. *Trends in Genetics* **12**, 192–196.

Cherry J.L., Young H., Di Sera L.J., Ferguson F.M., Kimball A.W., Dunn D.M., Gesteland R.F. and Weiss R.B. (1994) Enzyme-linked fluorescent detection for automated multiplex DNA sequencing. *Genomics* **20**, 68–74.

Chissoe S.L., Marra M.A., Hillier L., Brinkman R., Wilson R.K. and Waterson R.H. (1997) Representation of cloned genomic sequences in two sequencing vectors: correlation of DNA sequence and subclone distribution. *Nucleic Acids Research* **25**, 2960–2966.

Chowdhary B.P., Fronicke L., Gustavsson I. and Scherthan H. (1996) Comparative analysis of the cattle and human genomes: detection of ZOO-FISH and gene mapping-based chromosomal homologies. *Mammalian Genome* **7**, 297–302.

Chu G., Vollrath D. and Davis R. (1986) Separation of large DNA molecules by contour clamped homogenous electric fields. *Science* **234**, 1582–1585.

Chumakov I.M. and 14 others (1992a) Isolation of chromosome 21-specific yeast artificial chromosomes from a total human genome library. *Nature Genetics* **1**, 222–225.

Chumakov I. and 35 others (1992b) Continuum of overlapping clones spanning the entire human chromosome 21q. *Nature* **359**, 380–387.

Chumakov I.M. and 61 others (1995) A YAC contig of the human genome. *Nature* **377**, S175–S298.

Church G.M. and Kieffer-Higgins S. (1988) Multiplex DNA sequencing. *Science* **240**, 185–188.

Clarke L. and Carbon J. (1976) A colony bank containing synthetic Col E1 hybrid plasmids representative of the entire *E. coli* genome. *Cell* **9**, 91–99.

Clayton R.A., White O., Ketchum K.A. and Venter J.C. (1997) The first genome from the third domain of life. *Nature* **387**, 459–462.

Cole S.T. and Saint Girons I. (1994) Bacterial genomes. *FEMS Microbiology Reviews* **14**, 139–160.

Collins F. and Galas D. (1993) A new five-year plan for the U.S. human genome project. *Science* **262**, 43–46.

Collins F.S. (1992) Positional cloning: let's not call it reverse any more. *Nature Genetics* **1**, 3–6.

Collins F.S. (1995) Positional cloning moves from perditional to traditional. *Nature Genetics* **9**, 347–350.

Collins F.S., Drumm M.L., Cole J.L., Lockwood W.K., Van de Wonde G.F. and Iannuzzi M.C. (1987) Construction of a general human chromosome jumping library , with application to cystic fibrosis. *Science* **235**, 1046–1049.

Cooke R. and 33 others (1996) Further progress towards a catalogue of all *Arabidopsis* genes: analysis of a set of 5000 non-redundant ESTs. *Plant Journal* **9**, 101–124.

Copeland N.G. and 12 others (1993) A genetic linkage map of the mouse: current applications and future prospects. *Science* **262**, 57–66.

Cordell H.J. and Todd J.A. (1995) Multifactorial inheritance in type 1 diabetes. *Trends in Genetics* **11**, 499–504.

Cotton R.G.H. (1997) Slowly but surely towards better scanning for mutations. *Trends in Genetics* **13**, 43–46.

Coulson A. (1994) High-performance searching of biosequence databases. *Trends in Biotechnology* **12**, 76–80.

Coulson A., Sulston J., Brenner S. and Karn L. (1986) Toward a physical map of the genome of the nematode *Caenorhabditis elegans*. *Proceedings of the National Academy of Sciences, USA* **83**, 7821–7825.

Cox D.R., Burmeister M., Price E.R., Kim S. and Myers R.M. (1990) Radiation hybrid mapping: a somatic cell genetic method for constructing high-resolution maps of mammalian chromosomes. *Science* **250**, 245–250.

Cox D.R., Green E.D., Lander E.S., Cohen D. and Myers R.M. (1994) Assessing mapping progress in the human genome project. *Science* **265**, 2031–2032.

Craig J.M. and Bickmore W.A. (1994) The distribution of CpG islands in mammalian chromosomes. *Nature Genetics* **7**, 376–381.

Crawford A.M. and 11 others (1995) Sheep linkage mapping: nineteen linkage groups derived from the analysis of paternal half-sib families. *Genetics* **137**, 573–579.

Cross S.H. and Bird A.P. (1995) CpG islands and genes. *Current Opinion in Genetics and Development* **5**, 309–314.

D'Eustachio P. and Ruddle F.H. (1983) Somatic cell genetics and gene families. *Science* **220**, 919–928.

Daniels S.E. and 11 others (1996) A genome-wide search for quantitative trait loci underlying asthma. *Nature* **383**, 247–250.

Dausset J., Cann H., Cohen D., Lathrop M., Lalouel J.M. and White R.L. (1990) Collaborative genetic mapping of the Human Genome. *Genomics* **6**, 575–577.

Dauwerse J.G., Wiegant J., Raap A.K., Breuning M.H. and Van Ommen G.J.B. (1992) Multiple colors by fluorescence in situ hybridization using ratio labelled DNA probes create a molecular karyotype. *Human Molecular Genetics* **1**, 593–598.

Davidson E.H. and Britten R.J. (1973) Organization, transcription and regulation in the animal genome. *Quarterly Review of Biology* **48**, 565–613.

Davies K.E., Young B.D., Elles R.G., Hill M.E. and Williamson R. (1981) Cloning a representative genomic library of the human X chromosome after sorting by flow cytometry. *Nature* **293**, 374–376.

Dib C. and 13 others (1996) A comprehensive genetic map of the human genome based on 5,264 microsatellites. *Nature* **380**, 152–154.

Diehl S.R., Ziegle J., Buck G.A., Reynolds T.R. and Weber J.L. (1990) Automated genotyping of human DNA polymorphisms. *American Journal of Human Genetics* **47**, A177.

Dietrich W.F. and 14 others (1994) A genetic map of the mouse with 4006 simple sequence length polymorphisms. *Nature Genetics* **7**, 220–225.

Dietrich W.F. and 20 others (1996) A comprehensive map of the mouse genome. *Nature* **380**, 149–152.

Dimmock N.J. and Primrose S.B. (1994) *An Introduction to Modern Virology*. Blackwell Scientific Publications, Oxford.

Doebley J. (1992) Mapping the genes that made maize. *Trends in Genetics* **8**, 302–307.

Dogget N. (1992) In *The Human Genome Project*. Los Alamos Science No. 20. University Science Books, Sausalito, CA. 338 pp.

Donis-Keller H. and 32 others (1987) A genetic linkage map of the human genome. *Cell* **51**, 319–337.

Driscoll R.J., Youngquist M.G. and Baldeschweiler J.D. (1990) Atomic-scale imaging of DNA using scanning tunnelling microscopy. *Nature* **346**, 294–296.

Drmanac R., Drmanac S., Labat I., Crkvenjakov R., Vincentic A. and

References

Gemmel A. (1992) Sequencing by hybridization: towards an automated sequencing of one million M13 clones arrayed on membranes. *Electrophoresis* **13**, 566–573.

Drmanac R., Drmanac S., Strzoska Z., Paunesku T., Labat I., Zeremski M., Snoddy J., Funkhouser W.K., Koop B., Hood L. and Crkvenjakov R. (1993) DNA sequencing determination by hybridization: a strategy for efficient large scale sequencing. *Science* **260**, 1649–1652.

Drmanac R., Labat I., Brukner I. and Crkvenjakov R. (1989) Sequencing of megabase plus DNA by hybridization: theory of the method. *Genomics* **4**, 114–128.

Dujon B. (1996) The yeast genome project: what did we learn? *Trends in Genetics* **12**, 263–270.

Dujon B. and 107 others (1994) Complete DNA sequence of yeast chromosome XI. *Nature* **369**, 371–378.

Dujon B., Belfort M., Butow R.A., Jacq C., Lemieux C., Perlamma P.S. and Vost M. (1989) Mobile introns: definition of terms and recommended nomenclature. *Gene* **82**, 115–118.

Dunlapp D.D. and Bustamante C. (1989) Images of single-stranded nucleic acids by scanning tunnelling microscopy. *Nature* **342**, 204–206.

Dunn J.J. and Studier F.W. (1983) Complete nucleotide sequence of bacteriophage T7 DNA and the locations of T7 genetic elements. *Journal of Molecular Biology* **166**, 477–535.

Edwards A., Voss H., Rice P., Civitello A., Stegemann J., Schwager C., Zimmerman J., Erfle H., Caskey C.T. and Ansorge W. (1990) Automated DNA sequencing of the human HPRT locus. *Genome* **6**, 593–608.

Edwards M.D., Stuber C.W. and Wendel J.F. (1987) Molecular-marker-facilitated investigations of quantitative-trait loci in maize. I. Numbers, genomic distribution and types of gene action. *Genetics* **116**, 113–125.

Eggen A. and Fries R. (1995) An integrated cytogenetic and meiotic map of the bovine genome. *Animal Genetics* **26**, 215–236.

Elgar G, Sandford S., Macrae A., Venkatesh B. and Brenner S. (1996) Small is beautiful: comparative genomics with the pufferfish (*Fugu rubripes*). *Trends in Genetics* **12**, 145–150.

Eppig J.T. (1996) Comparative maps: adding pieces to the mammalian jigsaw puzzle. *Current Opinion in Genetics and Development* **6**, 723–730.

Evans G.A. and Lewis K.A. (1989) Physical mapping of complex genomes by cosmid multiplex analysis. *Proceedings of the National Academy of Sciences, USA* **86**, 5030–5034.

Fabret C., Quentin Y., Guiseppi A., Busuttil J., Haiech J. and Denizot F. (1995) Analysis of errors in finished DNA sequences: the surfactin operon of *Bacillus subtilis* as an example. *Microbiology* **141**, 345–350.

Fan J-B., Chikashige Y., Smith C.L., Niwa O., Yanagida M. and Cantor C.R. (1989) Construction of a *Not*I restriction map of the fission yeast *Schizosaccharomyces pombe* genome. *Nucleic Acids Research* **17**, 2801–2818.

Fauron C., Casper M., Gao Y. and Moore B. (1995) The maize mitochondrial genome: dynamic, yet functional. *Trends in Genetics* **11**, 228–235.

Felsenfeld A.L. (1996) Defining the boundaries of zebrafish developmental genetics. *Nature Genetics* **14**, 258–263.

Ferrin L.J. and Camerini-Otero R.D. (1991) Selective cleavage of human DNA: RecA-assisted restriction endonuclease (RARE) cleavage. *Science* **254**, 1494–1497.

Fickett J.W. (1996) Finding genes by computer: the state of the art. *Trends in Genetics* **12**, 316–320.

Fields C., Adams M.D., White O. and Venter J.C. (1994) How many genes in the human genome. *Nature Genetics* **7**, 345–346.

Fleischmann R.D. and 39 others (1995) Whole-genome random sequencing and assembly of *Haemophilus influenzae* Rd. *Science* **269**, 496–512.

Florijn R.J. and 9 others (1995) High resolution DNA fiberFISH genomic DNA mapping and colour barcoding of large genes. *Human Molecular Genetics* **4**, 831–836.

Fodor S.P.A., Rava R.P., Huang X.C., Pease A.C., Holmes C.P. and Adams C.L. (1993) Multiplexed biochemical assays with bioiogical-chips. *Nature* **364**, 555–556.

Fonstein M. and Haselkorn R. (1995) Physical mapping of bacterial genomes. *Journal of Bacteriology* **177**, 3361–3369.

Foote S., Vollrath D., Hilton A. and Page D.C. (1992) The human Y chromosome: overlapping DNA clones spanning the euchromatic region. *Science* **258**, 60–66.

Franco B. and 15 others (1991) A gene deleted in Kallman's syndrome shares homology with neural cell adhesion and axonal path-finding molecules. *Nature* **353**, 529–536.

Frankel W.N. (1995) Taking stock of complex trait genetics in mice. *Trends in Genetics* **11**, 471–477.

Fraser C.M. and 28 others (1995) The minimal gene complement of *Mycoplasma genitalium*. *Science* **270**, 397–403.

Fromont-Racine M., Rain J-C. and Legrain P. (1997) Towards a functional analysis of the yeast genome through exhaustive two-hybrid screens. *Nature Genetics* **16**, 277–281.

Fronicke L., Chowdhary B.P., Scherthan H. and Gustavsson I. (1996) Comparative map of the porcine and human genomes demonstrates ZOO-FISH and gene mapping-based chromosomal homologies. *Mammalian Genome* **7**, 285–290.

Fujii J., Otsu K., Zorgato F., De Leon S., Khanna V.K., Weiler J.E., O'Brien P.J. and MacLennan D.H. (1991) Identification of a mutation in porcine ryanodine receptor associated with malignant hyperthermia. *Science* **253**, 448–451.

Gardiner K. (1995) Human genome organization. *Current Opinion in Genetics and Development* **5**, 315–322.

Gelfand M.S. and Koonin E.V. (1997) Avoidance of palindromic words in bacterial and archaeal genomes: a close connection with restriction enzymes. *Nucleic Acids Research* **25**, 2430–2439.

Gerhold D. and Caskey C.T. (1997) It's the genes! EST access to human gene content. *BioEssays* **18**, 973–981.

Ghiso N.S., Parekh H. and Lennon G.G. (1993) A subset of 1200 hexamers is sufficient to sequence over 95% of cDNAs by hexamer string primer walking. *Genomics* **17**, 798–799.

Gilley J., Armes N. and Fried M. (1997) *Fugu* genome is not a good mammalian model. *Nature* **385**, 305–306.

Glazer A.N. and Mathies R.A. (1997) Energy-transfer fluorescent reagents for DNA analyses. *Current Opinion in Biotechnology* **8**, 94–102.

Goffeau A. (1997) Molecular fish on chips. *Nature* **385**, 202–203.

Goffeau A. and 15 others (1996) Life with 6000 genes. *Science* **274**, 546–567.

Goss S.J. and Harris H. (1975) New method for mapping genes in human chromosomes. *Nature* **255**, 680–684.

Goss S.J. and Harris H. (1977) Gene transfer by means of cell fusion II. The mapping of 8 loci on human chromosome 1 by statistical analysis of gene assortment in somatic cell hybrids. *Journal of Cell Science* **25**, 39–57.

Grandbastien M-A. (1992) Retroelements in higher plants. *Trends in Genetics* **8**, 103–108.

Green E.C. and Olson M.V. (1990) Chromosomal region of the cystic fibrosis gene in yeast artificial chromosomes: a model for human genome mapping. *Science* **250**, 94–98.

Gyapay G. and 13 others (1996) A radiation hybrid map of the human genome. *Human Molecular Genetics* **5**, 339–346.

Haley C.S., Knott S.A. and Elsen J-M. (1994) Mapping quantitative trait loci in crosses between outbred lines using least squares. *Genetics* **136**, 1195–1207.

Hanahan D. and Meselson M. (1980) Plasmid screening at high colony density. *Gene* **10**, 63–67.

Hancock J.M. (1996) Simple sequences in a 'minimal' genome. *Nature Genetics* **14**, 14–15.

Harrington J.J., Bokkelen G.V., Mays R.W., Gustashaw K. and Willard H.F. (1997) Formation of *de novo* centromeres and construction of first-generation human artificial microchromosomes. *Nature Genetics* **15**, 345–355.

Hartl D.L., Ajioka J.W., Cai H., Lohe A.R., Lovoskaya E.R., Smoller D.A. and Duncan I.W. (1992) Towards a *Drosophila* genome map. *Trends in Genetics* **8**, 70–75.

Hartl D.L., Nurminsky D.I., Jones R.W. and Lozovskaya E.R. (1994) Genome structure and evolution in *Drosophila* — applications of the framework P1 map. *Proceedings of the National Academy of Sciences* **91**, 6824–6829.

Hauge B.M., Hanley S., Giraudat J. and Goodman H.M. (1991) Mapping the *Arabidopsis* genome. In *Molecular Biology of Plant Development* (eds G. Jenkins and W. Schurch). Cambridge University Press, Cambridge.

Heale S.M., Stateva L.L. and Oliver S.G. (1994) Introduction of YACs into intact yeast cells by a procedure which shows low levels of recombinagenicity and co-transformation. *Nucleic Acids Research* **22**, 5011–5015.

Heiskanen M., Peltonen L. and Palotie A. (1996) Visual mapping by high resolution FISH. *Trends in Genetics* **12**, 379–382.

Hicks G.G., Shi E-G., Li X-M., Li C-H., Pawlak M. and Ruley H.E. (1997) Functional genomics in mice by tagged sequence mutagenesis. *Nature Genetics* **16**, 338–344.

Himmelreich R., Hilbert H., Plagens H., Pirkl E., Li B-C. and Herrmann R. (1996) Complete sequence analysis of the genome of the bacterium *Mycoplasma pneumoniae*. *Nucleic Acids Research* **24**, 4420–4449.

Himmelreich R., Plagens H., Hilbert H., Reiner B. and Herrmann R. (1997) Comparative analysis of the genomes of *Mycoplasma pneumoniae* and *Mycoplasma genitalium*. *Nucleic Acids Research* **25**, 701–712.

Hiratsuka J. and 15 others (1989) The complete sequence of the rice (*Oryza sativa*) chloroplast genome: inter-molecular recombination between distinct tRNA genes accounts for a major plastid DNA inversion during the evolution of the cereals. *Molecular and General Genetics* **217**, 185–194.

Hodgson J. (1994) Genome mapping the 'easy way'. *Biotechnology* **12**, 581–584.

Hoelzel R. (1990) The trouble with PCR machines. *Trends in Genetics* **6**, 237–238.

Hoffman E.P., Brown R.H. and Kunkel L.M. (1987) Dystrophin: the protein product of the Duchenne muscular dystrophy locus. *Cell* **51**, 919–928.

Hoheisel J.D. (1994) Application of hybridization techniques to genome mapping and sequencing. *Trends in Genetics* **10**, 79–83.

Hoheisel J.D., Maier E., Mott E., McCarthy L., Grigoriev A.V., Schalkwyk L.C., Nizetic D., Francis F. and Lehrach H. (1993) High resolution cosmid and P1 maps spanning the 14 Mb genome of the fission yeast *S. pombe*. *Cell* **73**, 109–120.

Holliday R. (1996) Endless quest. *BioEssays* **18**, 3–5.

Hudson T.J. and 50 others (1995) An STS-based map of the human genome. *Science* **270**, 1945–1954.

Hudson T.J., Colber A.M.E., Reeve M.P., Bae J.S., Lee M.K., Nussbaum R.L., Budarf M.L., Emanuel B.S. and Foote S. (1994). Isolation and regional mapping of 110 chromosome 22 STSs. *Genomics* **24**, 588–592.

Hughes A.L. and Hughes M.K. (1995) Small genomes for better flyers. *Nature* **377**, 391.

Ioannou P.A., Amemiya C.T., Garnes J., Kroisel P.M., Shizuya H., Chen C., Batzer M.A. and de Jong P.J. (1994) A new bacteriophage P1-derived vector for the propagation of large human DNA fragments. *Nature Genetics* **6**, 84–89.

Jacob H.J. and 19 others (1995) A genetic linkage map of the laboratory rat, *Rattus norvegicus*. *Nature Genetics* **9**, 63–69.

James M.R. and Lindpainter K. (1997) Why map the rat? *Trends in Genetics* **13**, 171–173.

James M.R. and 16 others (1994) A radiation hybrid map of 506 STS markers spanning human chromosome II. *Nature Genetics* **8**, 70–76.

Jasienski M. and Bazzaz F.A. (1995) Genome size and high CO_2. *Nature* **376**, 559–560.

Jasin M. (1996) Genetic manipulation of genomes with rare-cutting endonucleases. *Trends in Genetics* **12**, 224–228.

Jeffreys A.J. and Flavell R.A. (1977) The rabbit beta-globin gene contains a large insert in the coding squence. *Cell* **12**, 1097–1108.

Johnston M. (1996) Towards a complete understanding of how a simple eukaryotic cell works. *Trends in Genetics* **12**, 242–243.

Johnston M. and 34 others (1994) Complete nucleotide sequence of *Saccharomyces cerevisiae* chromosome VIII. *Science* **265**, 2077–2082.

Jordan B. (1993) *Travelling Around the Human Genome*. John Libbey Eurotext. 188 pp.

References

Jordan E. and Collins F.S. (1996) A march of genetic maps. *Nature* **380**, 11–12.

Kaneko T. and 23 others (1996) Sequence analysis of the genome of the unicellular cyanobacterium *Synechocystis* sp. strain PCC6803. II Sequence determination of the entire genome and assignment of potential protein coding regions. *DNA Research (Japan)* **3**, 185–209.

Khan A.S., Wilcox A.S., Polymeropoulos M.H., Hopkins J.A., Stevens T.J., Robinson M., Orpana A.K. and Sikela J.M. (1992) Single pass sequencing and physical and genetic mapping of human brain cDNAs. *Nature Genetics* **2**, 180–185.

Kieleczawa J., Dunn J.J. and Studier F.W. (1992) DNA sequencing by primer walking with strings of contiguous hexamers. *Science* **258**, 1787–1791.

Kinashi H., Shimaji M. and Sakai A. (1987) Giant linear plasmids in *Streptomyces* which code for antibiotic biosyntheis genes. *Nature* **328**, 454–456.

Kipling D. (1995) *The Telomere*. Oxford University Press, Oxford.

Ko M.S.H. (1990) An 'equalized cDNA library' by the reassociation of short double-stranded cDNAs. *Nucleic Acids Research* **18**, 5705–5711.

Kohara Y., Akiyama K. and Isono K. (1987) The physical map of the whole *E. coli* chromosome: application of a new strategy for rapid analysis and sorting of a large genomic library. *Cell* **50**, 495–508.

Konieczny A. and Ausubel F.A. (1993) A procedure for mapping *Arabidopsis* mutations using co-dominant ecotype-specific PCR-based markers. *The Plant Journal* **4**, 403–410.

Koob M. and Szbalski W. (1990) Cleaving yeast and *Escherichia coli* genomes at a single site. *Science* **1250**, 271–273.

Koonin E.V., Bork P. and Sander C. (1994) Yeast chromosome III: new gene functions. *EMBO Journal* **13**, 493–503.

Koonin E.V., Mushegian A.R. and Rudd K.E. (1996) Sequencing and analysis of bacterial genomes. *Current Biology* **6**, 404–416.

Koop B.F. (1995) Human and rodent DNA sequence comparisons: a mosaic model of genomic evolution. *Trends in Genetics* **11**, 367–371.

Kouprina N., Eldarov M., Moyzis R., Resnick M. and Larionov V. (1994) A model system to assess the integrity of mammalian YACs during transformation and propagation in yeast. *Genomics* **21**, 7–17.

Krishnan B.R., Jamry I., Berg D.E., Berg C.M. and Chaplin D.D. (1995) Construction of a genomic DNA 'feature map' by sequencing from nested deletions: application to the HLA class 1 region. *Nucleic Acids Research* **23**, 117–122.

Kumar A. (1996) The adventures of the Ty1-*copia* group of retrotransposons in plants. *Trends in Genetics* **12**, 41–43.

Kurata N. and 27 others (1994) A 300 kilobase interval genetic map of rice including 883 expressed sequences. *Nature Genetics* **8**, 365–372.

Laan M., Kallioniemi O-P., Hellsten E., Alitalo K., Peltonen L. and Palotie A. (1995) Mechanically stretched chromosomes as targets for high-resolution FISH mapping. *Genome Research* **5**, 13–20.

Lamb B.T. and Gearhart J.D. (1995) YAC transgenics and the study of genetics and human disease. *Current Opinion in Genetics and Development* **5**, 342–348.

Lamond A.I. (1988) RNA editing and the mysterious undercover genes of trypanosomatid mitochondria. *Trends in Biochemical Sciences* **13**, 283–284.

Lander E.S. (1996) The new genomics: global views of biology. *Science* **274**, 536–539.

Lander E.S. and Botstein D. (1987) Homozygosity mapping: a way to map human recessive traits with the DNA of inbred children. *Science* **236**, 1567–1570.

Lander E.S. and Botstein D. (1989) Mapping Mendelian factors underlying quantitative traits using RFLP linkage maps. *Genetics* **121**, 185–199.

Lander E.S. and Schork N.J. (1994) Genetic dissection of complex traits. *Science* **265**, 2037–2048.

Lander E.S. and Waterman M.S. (1988) Genomic mapping by fingerprinting random clones: a mathematical analysis. *Genomics* **4**, 231–239.

Larin Z., Monaco A.P. and Lehrach H. (1991) Yeast artificial chromosome libraries containing large inserts from mouse and human DNA. *Proceedings of the National Academy of Sciences, USA* **88**, 4123–4127.

Le Y. and Dobson M.J. (1997) Stabilization of yeast artificial chromosome clones in a *rad54-3* recombination-deficient host strain. *Nucleic Acids Research* **25**, 1248–1253.

Le Bourgeois P., Lautier M., Mara M. and Ritzenthaler P. (1992) New tools for the physical and genetic mapping of *Lactococcus* species. *Gene* **78**, 29–36.

Lee J-Y., Koi M., Stanbridge E.J., Oshimura M., Kumamoto A.T. and Feinberg A.P. (1994) Simple purification of human chromosomes to homogeneity using Muntjac hybrid cells. *Nature Genetics* **7**, 29–33.

Legouis R. and 14 others (1991) The candidate gene for the X-linked Kallman syndrome encodes a protein related to adhesion molecules. *Cell* **67**, 423–435.

Lehrach H., Drmanac R., Hoheisel J., Larin Z., Lennon G., Monaco A.P., Nizatic D., Zehetner G. and Poustka A. (1990) Hybridization fingerprinting in genome mapping and sequencing. In *Genome Analysis: Genetic and Physical Mapping* Volume 1, (eds K.E. Davies and S.M. Tishman), pp. 39–81. Cold Spring Harbor Laboratory Press.

Leipe D.D. (1996) Biodiversity, genomes, and DNA sequence databases. *Current Opinion in Genetics and Development* **6**, 686–691.

Lewin B. (1994) *Genes* V. Cell Press, Cambridge, MA/Oxford University Press, New York.

Lichter P., Chang C.J., Call K., Hermanson G., Evans G.A., Housman D. and Ward D. (1990) High-resolution mapping of human chromosome II by *in situ* hybridization with cosmid clones. *Science* **247**, 64–69.

Limbach P.A., Crain P.F. and McCloskey J.A. (1995) Characterization of oligonucleotides and nucleic acids by mass spectrometry. *Current Opinion in Biotechnology* **6**, 96–102.

Ling L.L., Ma N.S-F., Smith D.R., Miller D.D. and Moir D.T. (1993) Reduced occurrence of chimeric YACs in recombination-deficient hosts. *Nucleic Acids Research* **21**, 6045–6046.

Linger J., Cooper J.P. and Cech T.R. (1995) Telomerase and DNA end replication: no longer a lagging strand problem? *Science* **269**, 1533–1534.

References

Lisitsyn N.A. (1995) Representational difference analysis: finding the differences between genomes. *Trends in Genetics* **11**, 303–307.

Lisitsyn N., Lisitsyn N. and Wigler M. (1993) Cloning the difference between two complex genomes. *Science* **259**, 946–951.

Lisitsyn N.A., Segre J.A., Kusumi K., Lisitsyn N.M., Nadeau J.H., Frankel W.N., Wigler M.H. and Lander E.S. (1994) Direct isolation of polymorphic markers linked to a trait by genetically directed representational difference analysis. *Nature Genetics* **6**, 57–63.

Little P. (1996) Genome analysis. *Nature* **382**, 408.

Liu S-L., Hessel A. and Sanderson K.E. (1993) Genomic mapping with I-*Cen*I an intron-encoded endonuclease specific for genes for ribosomal RNA, in *Salmonella* spp., *Escherichia coli* and other bacteria. *Proceedings of the National Academy of Sciences, USA* **90**, 6874–6878.

Lizardi P.M. and Ward D.C. (1997) FISH with a twist. *Nature Genetics* **16**, 217–218.

Lovett M. (1994) Fishing for complements: finding genes by direct selection. *Trends in Genetics* **10**, 352–357.

Lovett M., Kere J. and Hinton L.M. (1991) Direct selection: a method for the isolation of cDNAs encoded by large genomic regions. *Proceedings of the National Academy of Sciences, USA* **88**, 9628–9633.

Ludecke H.J., Senger G., Claussen U. and Horsthemke B. (1989) Cloning defined regions of the human genome by microdissection of banded chromosomes and enzymatic amplification. *Nature* **338**, 348–350.

Lysov Y.P., Khorlin A.A., Khrapko K.R., Shick V.V., Florentiev V.L. and Mirzabekov A.D. (1988) DNA sequencing by hybridization with oligonucleotides: a novel method. *Proceedings of the USSR Academy of Sciences, USA* **303**, 1508–1511.

Maier E., Hoheisel J.D., McCarthy L., Mott R., Grigoriev A.V., Monaco A.P., Larin Z. and Lehrach H. (1992) Complete coverage of the *Schizosaccharomyces pombe* genome in yeast artificial chromosomes. *Nature Genetics* **1**, 273–277.

Makalowski W., Zhang J. and Boguski M.S. (1996) Comparative analysis of 1196 orthologous mouse and human full-length mRNA and protein sequences. *Genome Research* **6**, 846–857.

Mantegna R.N., Buldyrev S.V., Goldberger A.L., Havlin S., Peng C-K., Simons M. and Stanley H.E. (1994) Linguistic features of noncoding DNA sequences. *Physical Review Letters* **73**, 3169–3172.

Marahrens Y. and Stillman B. (1992) A yeast chromosomal origin of DNA replication defined by multiple functional elements. *Science* **255**, 817–823.

Marmur J., Rownd R. and Schildkraut C.L. (1963) Denaturation and renaturation of deoxyribonucleic acid. *Progress in Nucleic Acids Research* **1**, 231–300.

Martin G.B., de Vincente M.C. & Tanksley S.D. (1993) High-resolution linkage analysis and physical characterization of the *Pto* bacterial resistance locus in tomato. *Molecular Plant–Microbe Interactions* **6**, 26–34.

Mashal R.D., Koontz J. and Sklar J. (1995) Detection of mutations by cleavage of DNA heteroduplexes with bacteriophage resolvases. *Nature Genetics* **9**, 177–183.

178

Maxam A. and Gilbert W. (1977) A new method for sequencing DNA. *Proceedings of the National Academy of Sciences, USA* **74**, 560–564.

Mazur B.J. and Tingey S.V. (1995) Genetic mapping and introgression of genes of agronomic importance. *Current Opinion in Biotechnology* **6**, 175–182.

McCarthy L.C. (1996) Whole genome radiation hybrid mapping. *Trends in Genetics* **12**, 491–493.

McClelland M., Jones R., Patel Y. and Nelson M. (1987) Restriction endonucleases for pulsed field mapping of bacterial genomes. *Nucleic Acids Research* **15**, 5985–6005.

McCombie W.R., Adams M.D., Kelley J.M., FitzGerald M.G., Utterback T.R., Khan M., Dubnick M., Kerlavage A.R., Venter J.C. and Fields C. (1992a) *Caenorhabditis elegans* expressed sequence tags identify gene families and potential disease homologues. *Nature Genetics* **1**, 124–131.

McDonald J.F. (1993) Evolution and consequences of transposable elements. *Current Opinion in Genetics and Development* **3**, 855–864.

McFadden G.I., Gilson P.R., Douglas S.E., Cavalier-Smith T., Hofmann C.J.B. and Maier U-G. (1997) Bonsai genomics: sequencing the smallest eukaryotic genomes. *Trends in Genetics* **13**, 46–49.

Médigue C., Bouché J.P., Hénaut A. and Danchin A. (1990) Mapping of sequenced genes (1700 kbp) in the restriction map of the *Escherichia coli* chromosome. *Molecular Microbiology* **4**, 169–187.

Médigue C., Viari A., Hénaut A. and Danchin A. (1993) Colibri: a functional database for the *Escherichia coli* genome. *Microbiological Reviews* **57**, 623–654.

Meier-Ewert S., Maier E., Ahmodi A., Curtis J. and Lehrach H. (1993) An automated approach to generating expressed sequence catalogues. *Nature* **361**, 375–376.

Meng X., Benson K., Chada K., Huff E.J. and Schwartz D.C. (1995) Optical mapping of lambda bacteriophage clones using restriction endonucleases. *Nature Genetics* **9**, 432–438.

Mitas M. (1997) Trinucleotide repeats associated with human disease. *Nucleic Acids Research* **25**, 2245–2253.

Mizukami T. and 11 others (1993) A 13 kb resolution cosmid map of the 14 Mb fission yeast genome by non random sequence-tagged site mapping. *Cell* **73**, 121–132.

Monaco A.F. (1994) Isolation of genes from cloned DNA. *Current Opinion in Genetics and Development* **4**, 360–365.

Monaco A.P. and Larin Z. (1994) YACs, BACs, and MACs: artificial chromosomes as research tools. *Trends in Biotechnology* **12**, 280–286.

Monaco A.P., Neve R.L., Colletti-Feener C., Bertelson C.J., Kurnit D.M. and Kunkel L.M. (1986) Isolation of candidate cDNAs for portions of the Duchenne Muscular Dystrophy gene. *Nature* **323**, 646–650.

Monckton D.G. and Jeffreys A.J. (1993) DNA profiling. *Current Opinion in Biotechnology* **4**, 660–664.

Moore G., Devos K.M., Wang Z. and Gale M.D. (1995) Grasses, line up and form a circle. *Current Biology* **5**, 737–739.

Moore M.J. (1996) When the junk isn't junk. *Nature* **379**, 402–403.

Mott R., Grigoriev A., Maier E., Hoheisel J. and Lehrach H. (1993) Algorithms and software tools for ordering clone libraries: application to the mapping of the genome of *Schizosaccharomyces pombe*. *Nucleic Acids Research* **21**, 1965–1974.

Moyzis R.K., Torney E.C., Meyne J., Buckingham J.M., Wu J-R., Burks C., Sirotkin K.M. and Goad W.B. (1989) The distribution of interspersed repetitive DNA sequences in the human genome. *Genomics* **4**, 273–289.

Murray A.W. and Szostak J.W. (1983) Construction of artificial chromosomes in yeast. *Nature* **305**, 189–193.

Murray J.C. and 26 others (1994) A comprehensive human linkage map with centimorgan density. *Science* **265**, 2049–2054.

Nelson S.F., McCusker J.H., Sander M.A., Kee Y., Modrich P. and Brown P.O. (1993) Genomic mismatch scanning: a new approach to genetic linkage mapping. *Nature Genetics* **4**, 11–17.

Nillson M., Krejci K., Koch J., Kwiatkowski M., Gustavsson P. and Landegren U. (1997) Padlock probes reveal single-nucleotide differences, parent of origin and *in situ* distribution of centromeric sequences in human chromosomes 13 and 21. *Nature Genetics* **16**, 252–255.

Noolandi J., Forsyth J. and Shi A-C. (1993) Sequencing using pulsed field and image reconstruction. *Proceedings of the National Academy of Sciences* **90**, 10101–10104.

Ohyama K. and 12 others (1986) Chloroplast gene organization deduced from complete squence of liverwort *Marchantia polymorpha* chloroplast DNA. *Nature* **322**, 572–574.

Okubo K., Hori N., Matoba R., Niiyana T., Fukushima A., Kojima Y. and Matsubara K. (1992) Large scale cDNA sequencing for analysis of quantitative and qualitative aspects of gene expression. *Nature Genetics* **2**, 173–179.

Old R.W. and Primrose S.B. (1994) *Principles of Gene Manipulation*, 5th edn. Blackwell Scientific Publications, Oxford. 474 pp.

Oliver S.G. (1996a) From DNA sequence to biological function. *Nature* **379**, 597–600.

Oliver S. (1996b) A network approach to the systematic analysis of yeast gene function. *Trends in Genetics* **12**, 241–242.

Oliver S.G. and 119 others (1992) The complete DNA sequence of yeast chromosome III. *Nature* **357**, 38–46.

Olson M.V., Dutchik J.E., Graham M.Y., Brodeur G.M., Helms C., Frank M., MacCollin M., Scheinman R. and Frank T. (1986) A random-clone strategy for restriction mapping in yeast. *Proceedings of the National Academy of Sciences, USA* **83**, 7826–7830.

Palmer J.D. (1997) The mitochondrion that time forgot. *Nature* **387**, 454–455.

Pardue M.L., Danilevskaya O.N., Lowenhaupt K., Slot F. and Traverse K.L. (1996) *Drosophila* telomeres: new views on chromosome evolution. *Trends in Genetics* **12**, 48–52.

Parimoo S., Patanjali S.R., Shukle H., Chaplin D.D. and Weisman S.M. (1991) cDNA selection: efficient PCR approach for the selection of cDNAs encoded in large chromosomal DNA fragments. *Proceedings of the National Academy of Sciences, USA* **88**, 9623–9627.

Parra I. and Windle B. (1993) High resolution visual mapping of stretched DNA by fluorescent hybridization. *Nature Genetics* **5**, 17–21.

Paterson A.H. and 12 others (1996) Towards a unified genetic map of higher plants, transcending the monocot-dicot divergence. *Nature Genetics* **14**, 380–382.

Paterson A.H., Lander E.S., Hewitt J.D., Peterson S., Lincoln S.E. and Tanksley S.D. (1988) Resolution of quantitative traits into Mendelian factors by using a complete linkage map of restriction fragment length polymorphisms. *Nature* **335**, 721–726.

Pease A.C., Solas D., Sullivan E.J., Cronin M.T., Holmes C.P. and Fodor S.P.A. (1994) Light-generated oligonucleotide arrays for rapid DNA sequence analysis. *Proceedings of the National Academy of Science, USA* **91**, 5022–5026.

Pericak-Vance M.A. and Haines J.L. (1995) Genetic susceptibility to Alzheimer disease. *Trends in Genetics* **11**, 504–508.

Pierce J.C., Sauer B. and Sternberg N. (1992) A positive selection vector for cloning high molecular weight DNA by the bacteriophage P1 system: improved cloning efficiency. *Proceedings of the National Academy of Sciences, USA* **89**, 2056–2060.

Pinkel D., Straume T. and Gray J.W. (1986) Cytogenetic analysis using quantitative, high-sensitivity, fluorescence hybridization. *Proceedings of the National Academy of Sciences USA* **83**, 2934–2938.

Polymeropoulos M.H., Xiao H., Sikela J.M., Adams M., Venter J.C. and Merril C.R. (1993) Chromosomal distribution of 320 genes from a brain cDNA library. *Nature Genetics* **4**, 381–385.

Postlethwait J.H. and 14 others (1994) A genetic linkage map for the zebrafish. *Science* **264**, 699–704.

Postlethwait J.H. and Talbot W.S. (1997) Zebrafish genomics: from mutations to genes. *Trends in Genetics* **13**, 183–190.

Poustka A., Rohl T.M., Barlow D.P., Frischauf A.M. and Lehrach H. (1987) Construction and use of human chromosome jumping libraries from *Not*I-digested DNA. *Nature* **325**, 353–355.

Prescott D.M. (1994) The DNA of ciliated protozoa. *Microbiological Reviews* **58**, 233–267.

Quesada M.A. (1997) Replaceable polymers in DNA sequencing by capillary electrophoresis. *Current Opinion in Biotechnology* **8**, 82–93.

Ramirez-Solis R., Liu P. and Bradley A. (1995) Chromosome engineering in mice. *Nature* **378**, 720–724.

Reed P.W. and 13 others (1994) Chromosome-specific microsatellite sets for fluorescence-based, semi-automated genome mapping. *Nature Genetics* **7**, 390–395.

Reiter R.S., Williams J.G.K., Feldmann K.A., Rafalsta A., Tingey S.V. and Scolnick P.A. (1992) Global and local genome mapping in *Arabidopsis thaliana* by using recombinant inbred lines and random amplified polymorphic DNAs. *Proceedings of the National Academy of Sciences, USA* **89**, 1477–1481.

Richards J.E., Gilliam T.C., Cole J.L., Drumm M.L., Wasmuth J.J., Gusella J.F. and Collins F.S. (1988) Chromosome jumping from DS410(G8) toward the Huntington disease gene. *Proceedings of the National Academy of Sciences, USA* **85**, 6437–6441.

Riley M. (1993) Functions of the gene products of *Escherichia coli*. *Microbiological Reviews* **57**, 862–952.

Robinson K., Gilbert W. and Church G.M. (1994) Large scale bacterial gene discovery by similarity search. *Nature Genetics* **7**, 205–214.

References

Rommens J.M. and 14 others (1989) Identification of the cystic fibrosis gene: chromosome walking and jumping. *Science* **245**, 1059–1065.

Rosenberg C., Florijn R.J., Van De Rijke F.M., Blonden L.A.J., Raap T.K., Van Ommen G-J.B. and Den Dunnen J.T. (1995) High resolution DNA fiber-fish on yeast artificial chromosomes: direct visualization of DNA replication. *Nature Genetics* **10**, 477–479.

Rosenberg M., Przybylska M. and Straus D. (1994) RFLP subtraction: a method for making libraries of polymorphic markers. *Proceedings of the National Academy of Sciences* **91**, 6113–6117.

Ross M.T. and 10 others (1992) Selection of a human chromosome 21 enriched YAC sub-library using a chromosome-specific composite probe. *Nature Genetics* **1**, 284–290.

Roush W. (1997) A zebrafish genome project? *Science* **275**, 923.

Rubattu S., Volpe M., Kreutz R., Ganten U., Ganten D. and Lindpainter K. (1996) Chromosomal mapping of quantitative trait loci contributing to stroke in a rat model of complex human disease. *Nature Genetics* **13**, 429–434.

Rudd K.E., Miller W., Ostell J. and Benson D.A. (1990) Alignment of *Escherichia coli* K12 DNA sequences to a genomic restriction map. *Nucleic Acids Research* **18**, 313–321.

Rudd K.E., Miller W., Werner C., Ostell J., Toltoshev C. and Satterfield S.G. (1991) Mapping sequenced *E. coli* genes by computer: software, strategies and examples. *Nucleic Acids Research* **19**, 637–647.

Saitoh Y. and Laemmli U.K. (1994) Metaphase chromosome structure: bands arise from a differential folding path of the highly AT-rich scaffold. *Cell* **76**, 609–622.

Samad A.H. and 19 others (1995) Mapping the genome one molecule at a time — optical mapping. *Nature* **378**, 516–517.

Sanger F., Air G.M., Barrell B.G., Brown N.L., Coulson A.R., Fiddes J.C., Hutchinson C.A., Slocombe P.M. and Smith M. (1977a) Nucleotide sequence of bacteriophage ΦX174DNA. *Nature* **265**, 687–695.

Sanger F., Nicklen S. and Coulson A.R. (1977b) DNA sequencing with chain terminating inhibitors. *Proceedings of the National Academy of Sciences, USA* **74**, 5463–5467.

Sanger F., Coulson A.R., Hong G.F., Hill D.F. and Petersen G.B. (1982) Nucleotide sequence of bacteriophage λ DNA. *Journal of Molecular Biology* **162**, 729–773.

Sargent C.A. and 10 others (1993) Cloning of the X-linked glycerol kinase deficiency gene and its identification by sequence comparison to the *Bacillus subtilis* homologue. *Human Molecular Genetics* **2**, 97–106.

Sasaki T. and 20 others (1994) Toward cataloguing all rice genes: large-scale sequencing of randomly chosen rice cDNAs from a callus cDNA library. *Plant Journal* **6**, 615–624.

Sassaman D.M. and 9 others (1997) Many human L1 elements are capable of retrotransposition. *Nature Genetics* **16**, 37–43.

Schuler G.D. and 103 others (1996). A gene map of the human genome. *Science* **274**, 640–546.

Schumacher A., Faust C. and Magnuson T. (1996) Positional cloning of a global regulator of anterior-posterior patterning in mice. *Nature* **383**, 250–253.

Schwartz D.C. and Cantor C.R. (1984) Separation of yeast chromosome-sized DNAs by pulsed field gradient gel electrophoresis. *Cell* **37**, 67–75.

Schwartz D.C., Li X., Hernandez L.I., Ramnarian S.P., Huff E.J. and Wang Y-K (1993) Ordered restriction maps of *Sacharomyces cerevisiae* chromosomes constructed by optical mapping. *Science* **202**, 110–114.

Sedlacek Z., Konecki D.S., Siebenhaar R., Kioschis P. and Poustka A. (1993) Direct selection of DNA conserved between species. *Nucleic Acids Research* **21**, 3419–3425.

Serikawa T. and 10 others (1992) Rat gene mapping using PCR-analyzed microsatellites. *Genetics* **131**, 701–721.

Sheffield V.C., Nishimura D.Y. and Stone E.M. (1995) Novel approaches to linkage mapping. *Current Opinion in Genetics and Development* **5**, 335–341.

Shields R. (1996) Plant chromosomes unite. *Trends in Genetics* **12**, 37–38.

Shimamura M. and 8 others (1997) Molecular evidence from retroposons that whales form a glade within even-toed ungulates. *Nature* **388**, 666–669.

Shinozaki K. and 22 others (1986) The complete nucleotide sequence of the tobacco chloroplast genome: its gene organization and expression. *EMBO Journal* **5**, 2043–2049.

Shippen D.E. (1993) Telomeres and telomerases. *Current Opinion in Genetics and Development* **3**, 759–763.

Shizuya H., Birren B., Kim U-J., Mancino V., Slepak T., Tachiiri Y. and Simon M. (1992) Cloning and stable maintenance of 300-kilobase-pair fragments of human DNA in *Escherichia coli* using an F-factor-based vector. *Proceedings of the National Academy of Sciences, USA* **89**, 8794–8797.

Shoemaker D.D., Lashkari D.A., Morris D., Mittmann M. and Davis R.W. (1996) Quantitative phenotypic analysis of yeast deletion mutants using a highly parallel molecular bar-coding strategy. *Nature Genetics* **14**, 450–460.

Sidén-Kiamos I., Saunders R.D.C., Spanos L., Majerus T., Treanear J., Savakas C., Louis C., Glover D.M., Ashburner M. and Kafatos F.C. (1990) Towards a physical map of the *Drosophila melanogaster* genome: mapping of cosmid clones within defined genomic divisions. *Nucleic Acids Research* **18**, 6261–6270.

Sikela J.M. and Auffray C. (1993) Finding new genes faster than ever. *Nature Genetics* **3**, 189–191.

Singer M. and Berg P. (1990) *Genes and Genomes*. Blackwell Scientific Publications, Oxford.

Smith C.L., Econome J.G., Schutt A., Klco S. and Cantor C.R. (1987) A physical map of the *Escherichia coli* K12 genome. *Science* **236**, 1448–1453.

Smith M.W., Holmsen A.L., Wei Y.H., Peterson M. and Evans G.A. (1994) Genomic sequence sampling: a strategy for high resolution sequence-based physical mapping of complex genomes. *Nature Genetics* **7**, 40–47.

Smith V., Botstein D. and Brown P.O. (1995) Genetic footprinting: a genomic strategy for determining a gene's function given its sequence. *Proceedings of the National Academy of Sciences* **92**, 6479–6483.

References

Sofia H.J., Burland V., Daniels D.L., Plunkett G. and Blattner F.R. (1994) Analysis of the *Escherichia coli* genome. V. DNA sequence of the region from 76.0 to 81.5 minutes. *Nuceic Acids Research* **22**, 2576–2586.

Song W-Y. and 11 others (1995) A receptor kinase-like protein encoded by the rice disease resistance gene, Xa21. *Science* **270**, 1804–1806.

Southern E. (1996) DNA chips: analysing sequence by hybridization to oligonucleotides on a large scale. *Trends in Genetics* **12**, 110–115.

Southern E.M. (1988) Analyzing polynucleotide sequences. *International Patent Application PCT GB 89/01114*.

Southern E.M., Maskos U. and Elder J.K. (1992) Analyzing and comparing nucleic acid sequences by hybridization to arrays of oligonucleotides: evaluation using experimental models. *Genomics* **13**, 1008–1017.

Sternberg N. (1990) Bacteriophage P1 cloning system for the isolation, amplification and recovery of DNA fragments as large as 100 kilobase pairs. *Proceedings of the National Academy of Sciences, USA* **87**, 103–107.

Sternberg N. (1994) The P1 cloning system — past and future. *Mammalian Genome* **5**, 397–404.

Stoesser G., Sterk P., Tuli M.A., Stoehr P.J. and Cameron G.N. (1997) The EMBL nucleotide sequence database. *Nucleic Acids Research* **25**, 7–13.

Strachan T., Arbitol M, Davidson D. and Beckmann J.J. (1997). A new dimension for the human genome project: towards comprehensive expression maps. *Nature Genetics* **16**, 126–132.

Sulston J., Mallet F., Staden R., Durbin R., Horsnell T. and Coulson A. (1988) Software for genome mapping by fingerprint techniques. *CABIOS* **4**, 125–132.

Sun T-Q., Fernstermacher D.A. and Ves J-M.H. (1994) Human artificial episomal chromosomes for cloning large DNA fragments in human cells. *Nature Genetics* **8**, 33–41.

Sutherland G.R. and Richards R.I. (1995). The molecular basis of fragile sites in human chromosomes. *Current Opinion in Genetics and Development* **5**, 323–327.

Szybalski W. (1997) RecA-mediated Achilles heel cleavage. *Current Opinion in Biotechnology* **8**, 75–81.

Tabor S. and Richardson C.C. (1995) A single residue in DNA polymerases of the *Escherichia coli* DNA polymerase I family is critical for distinguishing between deoxy- and dideoxyribonucleotides. *Proceedings of the National Academy of Sciences* **92**, 6339–6343.

Tanksley S.D., Ganal M.W. and Martin G.B. (1995) Chromosome landing: a paradigm for map-based gene clones in plants with large genomes. *Trends in Genetics* **11**, 63–68.

Tateno Y. and Gojobori T. (1997) DNA databank of Japan in the age of information biology. *Nucleic Acids Research* **25**, 14–17.

Tatusov R.L., Mushegian A.R., Bork P., Brown N.P., Hayes W.S., Borodovsky M., Rudd K.E. and Koonin E.V. (1996) Metabolism and evolution of Haemophilus influenzae deduced from a whole-genome comparison with *Escherichia coli*. *Current Biology* **6**, 279–291.

Thierry A. and Dujon B. (1992) Nested chromosomal fragmentation in yeast using the meganuclease I-*Sce*I: a new method for physical mapping of eukaryotic genomes. *Nucleic Acids Research* **20**, 5625–5631.

184

Tijssen P. (1993) *Hybridization with Nucleic Acid Probes. Part 1: Theory and Nucleic Acid Preparation.* Elsevier, Amsterdam.

Tingey S.V. and Del Tufo J.P. (1993) Genetic analysis with RAPD markers. *Plant Physiology* **101**, 349–352.

Tomb J-F., and 41 others (1997) The complete genome sequence of the gastric pathogen *Helicobacter pylori*. *Nature* **388**, 539–547.

Trask B., Christensen M., Fertitta A., Bergmann A., Ashworth L., Branscomb E., Carrano A. and Van den Engh G. (1992) Fluorescence *in situ* hybridization mapping of human chromosome 19: mapping and verification of cosmid contigs formed by random restriction finger-printing. *Genomics* **14**, 162–167.

Trask B.J., Pinkel D and Van den Engh G.J. (1989) The proximity of DNA sequences in interphase cell nuclei is correlated to genomic distance and permits ordering of cosmids spanning 250 kilobase pairs. *Genomics* **5**, 710–717.

Travis G.H. and Sutcliffe J.G. (1988) Phenol emulsion-enhanced DNA-driven subtractive cDNA cloning: isolation of low-abundance monkey cortex-specific mRNAs. *Proceedings of the National Academy of Sciences, USA* **85**, 1696–1700.

Trower M.K. and 13 others (1996) Conservation of synteny between the genome of the pufferfish (*Fugu rubripes*) and the region on human chromosome 14 (14q24.3) associated with familial Alzheimer disease (*AD3* locus). *Proceedings of the National Academy of Sciences* **93**, 1366–1369.

Tsai J-Y., Namin-Gonzalez M.L. and Silver L.M. (1994) False association of human ESTs. *Nature Genetics* **8**, 321–322.

Tyler-Smith C. and Willard H.F. (1993) Mammalian chromosome struc-ture. *Current Opinion in Genetics and Development* **3**, 390–397.

Unseld M., Marienfeld J.R., Brandt P. and Brennicke A. (1997) The mitochondrial genome of *Arabidopsis thaliana* contains 57 genes in 366,924 nucleotides. *Nature Genetics* **15**, 57–61.

Van Ommen G-B., Breuning M.H. and Raap A.K. (1995) FISH in genome research and molecular diagnostics. *Current Opinion in Genetics and Development* **5**, 304–308.

Venter J.C., Smith H.O. and Hood L. (1996) A new strategy for genome sequencing. *Nature* **381**, 364–366.

Von Haeseler A., Sajantila A. and Paabo S. (1995) The genetic archaeol-ogy of the human genome. *Nature Genetics* **14**, 135–140.

Vos P. and 10 others (1995) AFLP: a new technique for DNA finger-printing. *Nucleic Acids Research* **23**, 4407–4414.

Voss H., Schwager C., Wiemann S., Zimmermann J., Stegemann J., Erfle H., Voie A.M., Drzonek H. and Ansorge W. (1995) Efficient low redundancy large-scale sequencing at EMBL. *Journal of Biotechnology* **41**, 121–129.

Voytas D.F. (1996) Retroelements in genome organization. *Science* **274**, 737–738.

Wada M., Little R.D., Abidi F., Porta G., Labella T., Cooper T., Delle Valle G., D'Urso M. and Schlessinger D. (1990) Human Xq24-Xq28: approaches to mapping with yeast artificial chromosomes. *American Journal of Human Genetics* **46**, 95–105.

Walter M.A., Spillett D.J., Thomas P., Weissenbach J. and Goodfellow

References

P.N. (1994) A method for constructing radiation hybrid maps of whole genomes. *Nature Genetics* 7, 22–28.

Wan K-I., Blackwell J.M. and Ajioka J.W. (1995). *Toxoplasma gondii* expressed sequence tags: insite into tachyzoite gene expression. *Molecular Biochemical Parasitology* 75, 179–186.

Wang G-L., Holsten T.E., Song W-Y., Wang H-P. and Ronald P.C. (1995a) Construction of a rice bacterial artificial chromosome library and identification of clones linked to the Xa-21 disease resistance locus. *The Plant Journal* 7, 525–533.

Wang Y-K., Huff E.J. and Schwartz D.C. (1995b) Optical mapping of site-directed cleavages on single DNA molecules by the RecA-assisted restriction endonuclease technique. *Proceedings of the National Academy of Sciences* 92, 165–196.

Wang Y-K., Prade R.A., Griffith J., Timberlake W.E. and Arnold J. (1994) A fast random cost algorithm for physical mapping. *Proceedings of the National Academy of Sciences* 91, 11094–11098.

Warren S.T. (1996) The expanding world of trinucleotide repeats. *Science* 271, 1374–1375.

Waterston R. and 17 others (1992) A survey of expressed genes in *Caenorhabditis elegans*. *Nature Genetics* 1, 114–123.

Weinstock K.G., Kirkness E.F., Lee N.H., Earle-Hughes J.A. and Venter J.C. (1994) cDNA sequencing: a means of understanding cellular physiology. *Current Opinion in Biotechnology* 5, 599–603.

Weissenbach J., Gyapay G., Dib C., Vignal A., Morissette J., Millasseau P., Vaysséix G. and Lathrop M. (1992) A second generation linkage map of the human genome based on highly informative microsatellite loci. *Nature* 359, 794–802.

Williams J.G.K., Kubelik A.R., Livak K.J., Rafalski J.A. and Tingey S.V. (1990) DNA polymorphisms amplified by arbitrary primers are useful as genetic markers. *Nucleic Acids Research* 18, 6531–6535.

Williams S.M. and Robbins L.G. (1992) Molecular genetic analysis of *Drosophila* rDNA arrays. *Trends in Genetics* 8, 335–340.

Wilson R. and 54 others (1994) 2.2 Mb of contiguous nucleotide sequence from chromosome III of *C. elegans*. *Nature* 358, 32–38.

Wilson R.K., Koop B.F., Chen C., Halloran N., Sciammis R. and Hood L. (1992) Nucleotide sequence analysis of 95 kb near the 3′ end of the murine T-cell receptor α/δ chain locus: strategy and methodology. *Genomics* 13, 1198–1208.

Woo S-S., Jiang J., Gill B.S., Paterson A.H. and Wing R.A. (1994) Construction and characterisation of a bacterial artificial chromosome library of *Sorghum bicolor*. *Nucleic Acids Research* 22, 4922–4931.

Woolley A.T. and Mathies R.A. (1995) Ultra-high-speed DNA sequencing using capillary electrophoresis chips. *Analytical Chemistry* 67, 3676–3680.

Wu R. and Taylor E. (1971) Nucleotide sequence analysis of DNA. II. Complete nucleotide sequence of the cohesive ends of bacteriophage λ DNA. *Journal Molecular Biology* 57, 491–511.

Wynshaw-Boris A. (1996) Model mice and human disease. *Nature Genetics* 13, 259–260.

Yager T.D., Dunn J.M. and Stevens J.K. (1997) High-speed DNA sequencing in ultrathin slab gels. *Current Opinion in Biotechnology* **8**, 107–113.

Yahraus T., Braverman N., Dodt G., Kalish J.E., Morrell J.C., Moser H.W., Valle D. and Gould S.J. (1996) The peroxisome biogenesis disorder group 4 gene, *PXAAA1*, encodes a cytoplasmic ATPase required for stability of the PTS1 receptor. *EMBO Journal* **15**, 2914–2923.

Yokota H. and 9 others (1997) A new method for straightening DNA molecules for optical restriction mapping. *Nucleic Acids Research* **25**, 1064–1070.

Yokota H., Van Den Engh G., Hearst J.E., Sachs R.K. and Trask B.J (1995) Evidence for the organization of chromatin in megabase pair-sized loops arranged along a random walk path in the human G0/G1 interphase nucleus. *Journal of Cell Biology* **130**, 1239–1249.

Youil R., Kemper B.W. and Cotton R.G.H. (1995) Screening for mutations by enzyme mismatch cleavage with T4 endonuclease VII. *Proceedings of the National Academy of Sciences* **92**, 87–91.

Yung J-F. (1996) New FISH probes — the end in sight. *Nature Genetics* **14**, 10–12.

Zehetner G. and Lehrach H. (1994) The reference library system — sharing biological material and experimental data. *Nature* **367**, 489–491.

Zhang P., Schon E.A., Fischer S.G., Cayanis E., Weiss J., Kistler S. and Bourne P.E. (1994) An algorithm based on graph theory for the assembly of contigs in physical mapping of DNA. *CABIOS* **10**, 309–317.

Zorio D.A.R., Cheng N.N., Blumenthal T. and Spieth J. (1994) Operons as a common form of chromosomal organization in C. *elegans*. *Nature* **372**, 270–272.

Index